SpringerBriefs in Mathematics

SpringerBriefs in Mathematics showcase expositions in all areas of mathematics and applied mathematics. Manuscripts presenting new results or a single new result in a classical field, new field, or an emerging topic, applications, or bridges between new results and already published works, are encouraged. The series is intended for mathematicians and applied mathematicians.

More information about this series at http://www.springer.com/series/10030

SpringerBriefs present concise summaries of cutting-edge research and practical applications across a wide spectrum of fields. Featuring compact volumes of 50 to 125 pages, the series covers a range of content from professional to academic. Briefs are characterized by fast, global electronic dissemination, standard publishing contracts, standardized manuscript preparation and formatting guidelines, and expedited production schedules.

Typical topics might include:

- A timely report of state-of-the art techniques
- A bridge between new research results, as published in journal articles, and a contextual literature review
- A snapshot of a hot or emerging topic
- An in-depth case study
- A presentation of core concepts that students must understand in order to make independent contributions

Titles from this series are indexed by Web of Science, Mathematical Reviews, and zbMATH.

Nicoló Gusmeroli

Machine Scheduling to Minimize Weighted Completion Times

The Use of the α-point

Springer

Nicoló Gusmeroli
Institut für Mathematik
Alpen-Adria-Universität Klagenfurt
Klagenfurt, Kärnten
Austria

ISSN 2191-8198 ISSN 2191-8201 (electronic)
SpringerBriefs in Mathematics
ISBN 978-3-319-77527-2 ISBN 978-3-319-77528-9 (eBook)
https://doi.org/10.1007/978-3-319-77528-9

Library of Congress Control Number: 2018935956

Printed on acid-free paper

This Springer imprint is published by the registered company Springer International Publishing AG part of Springer Nature
The registered company address is: Gewerbestrasse 11, 6330 Cham, Switzerland

To those who ever believed in me…

Preface

This work reviews the most important results regarding the use of the α-point in Scheduling Theory. It provides a number of different LP relaxations for scheduling problems and seeks to explain their polyhedral consequences. It also explains the concept of the α-point and how the conversion algorithm works, pointing out the relations to the sum of the weighted completion times. Lastly, the book explores the latest techniques used for many scheduling problems with different constraints, such as release dates, precedences, and parallel machines.

This reference book is intended for advanced undergraduate and postgraduate students who are interested in Scheduling Theory. It is also inspiring for researchers wanting to learn about sophisticated techniques and open problems of the field.

Klagenfurt, Austria — Nicoló Gusmeroli
January 2018

Contents

About the Author

Nicoló Gusmeroli completed his master's degree at the ELTE University of Budapest in 2017 and is currently working on the project High-Performance Solver for Binary Quadratic Problems at the Alpen-Adria University of Klagenfurt as a Ph.D. student. His main research interests are in combinatorial optimization, semidefinite optimization, and Scheduling Theory.

He completed his bachelor's studies at the University of Verona prior to spending an exchange semester at the University of Primorska (Slovenia), where he wrote his bachelor's thesis.

Chapter 1
Machine Scheduling to Minimize Weighted Completion Times

1 Introduction

Scheduling Theory is a branch of Mathematics, more precisely of Operations Research, which deals with finding the best allocation of scarce resources to activities over time. More generally we say that scheduling problems involve jobs and machines, and we want to find the optimal ordering of the jobs over the machines in such a way to minimize (or maximize) an objective function. Obviously there can be some constraints on the possible orderings, so we want to find a schedule specifying when and on which machine every job must be processed. For a precise definition, see the Encyclopedia of Mathematics, [20].

In all our problems, we have a set of jobs $N = \{j_1, j_2, \ldots, j_n\}$, and each of them has associated a processing time p_j. Then, following the notation in [17], we can define a scheduling problem by the three field notations: $\alpha|\beta|\gamma$, where α is the machine environment, β represents the set of constraints and characteristics, and γ denotes the optimization criterion. We also assume that any machine can process at most one job at any time and any job can be processed simultaneously on at most one machine.

The machine environment specifies the number of machines and, in case of more than one machines, the possible difference of speed between them. The easiest case is with a single machine, and it will be our basis case. In addition, in Sect. 7, we will introduce the concept of machines with different speeds, so if m_1 has normal speed and m_2 has double speed, we have that job j takes p_j times to be processed on machine m_1 and $\frac{p_j}{2}$ if it is processed on machine m_2. Then, in Sect. 8, we will use several machines that are identical from the view of point of the jobs, so we say that these are m parallel machines.

N. Gusmeroli, *Machine Scheduling to Minimize Weighted Completion Times*, SpringerBriefs in Mathematics, https://doi.org/10.1007/978-3-319-77528-9_1

The set of constraints and characteristics can be various, so we will introduce just the ones we will use:

- *pmtn*: this means that the job can be preempted, so it can be started on one machine, stopped, and resumed in a later time, also on a different machine;
- r_j: this indicates the presence of some release dates, so if for job j we have $r_j = x$ it means that job j can be started only after x units of time;
- *prec*: this means that there is a partial order between the jobs, so if, for jobs j and k, $j \prec k$, then job j must be completed before starting job k;
- d_{ij}: this is a more generic parameter and can generalize both r_j and *prec*. The set of d_{ij} defines a partial order on the jobs and we have that, for two jobs j and k, d_{jk} means that job k can start only when d_{jk} units of time passed after the completion of job j.

However, all the constraints used will be better explained.

The optimality criterion depends on the problem. Let us define C_j^S as the completion time of job j in schedule S. Then the usually used criteria are the minimization of $\sum C_j$, so the sum of completion times, and, if we associate to each job j a weight w_j, the minimization of the weighted sum of completion time, so $\min \sum w_j C_j$ (we can consider the former one as the latter with w identically 1, $w \equiv 1$). These two criteria could be seen also as the average completion and the average weighted completion, respectively, since we have just to divide by the number of jobs.

Some scheduling problems can be solved optimally, like $1||\sum C_j$, by scheduling jobs by a shortest processing time first rule, $1||\sum w_j C_j$ and the unit processing time problem $1|r_j, p_j = 1|\sum w_j C_j$, by using the Smith's rule that is scheduling jobs by nonincreasing $\frac{w_j}{p_j}$ (see [34, 42]), and $1|r_j, pmtn|\sum C_j$, scheduling the jobs with a shortest remaining processing time first rule (see [3, 22]).

Unfortunately, with precedence constraints, release dates or parallel machines, scheduling problems turn out to be (strongly) NP-hard, by example $1|r_j|\sum w_j C_j$ is NP-hard, also for $w \equiv 1$, see [28]. In such cases we can just search for approximate solutions: a ρ-approximation algorithm is a polynomial time algorithm guaranteed to deliver a solution of cost at most ρ times the optimum value; we refer to [18] for a survey.

One of the key ingredients in the design of approximation algorithms is the choice of a bound on the optimum value.

In Sect. 3 we will see some linear programming-based lower bounds. Dyer and Wolsey, in [13], formulated two time-indexed relaxations, one preemptive and one nonpreemptive. They showed that the nonpreemptive one is as least as good as the preemptive, but they could not say that it is always better. We will see the preemptive time-indexed relaxation, and we will compare it with a completion time relaxation, proposed first in [37]. By the use of the shifted parallel inequalities, firstly presented in [32], we will prove the equality between them, see [14]. In [13] the authors proved the polynomial time solvability of the preemptive time-indexed relaxation since it turns out to be a transportation problem, hence it can be solved in polynomial time, despite its exponential size. In [45] these formulations were proved to be equal also to the combinatorial lower bounds given in [6], that is based on job splitting. We will also prove the supermodularity of these functions, in order to see, as pointed in [34], that we can find the LP schedule from a greedy algorithm. Other important consequences following the supermodularity can be seen in [33, 34].

In Sect. 4 we introduce the concept of α-point of a job, defined the first time in [30]. In the same paper was described an algorithm that converts a preemptive schedule to a nonpreemptive schedule, this conversion algorithm is called α-CONV and it plays a crucial role in the study of the algorithms, since it delays the start of the jobs in such a way to relate them with the LP formulation. In order to have a detailed discussion of the use of the α-points, it is possible to see [40].

In Sect. 5, we will make use of the α-conversion algorithm. We will use also the concept of canonical decomposition, introduced in [14] in order to obtain better bounds. We will present first the bound given by using a fixed α and then an improved result for a uniformly random chosen α for $1|r_j|\sum w_jC_j$, proved in [15]. Then we will show some improvements, presenting the current best bounds given in [16], by using some specific random α distribution and the concept of conditional probabilities, see [29].

In Sect. 6 we study the problem $1|r_j|\sum C_j$, so the particular case of the problem studied in Sect. 5 with the restriction that we have, now, identical weights. Following the ideas in [26, 30] we use the concept of α-point in a different manner in order to present the job-by-job bound proved by Chekuri et al. in [10] that improves the results of the previous sections. We also present a family of instances for this problem for which our LP-based algorithm reaches the approximation guaranteed, so the bound, for this particular LP relaxation, is tight. Hence if we want to improve this result, we have to study some different relaxation as basis.

In Sect. 7, we study the problem $1|r_j, prec|\sum w_jC_j$. We will show the recent result given by Skutella in [41] that improves the previous best approximation ratios, see [36, 37]. In order to prove the result the author has little strengthened the 2-approximation of Hall et al. see [19], for $1|r_j, prec, pmtn|\sum w_jC_j$, showing that scheduling in order of increasing LP completion times on a double speed machine yields a schedule whose cost is at most the cost of an optimal schedule on a regular machine. He modified the analysis studied by Chekuri et al. in [10] showing how to turn the preemptive schedule on a double speed machine into a nonpreemptive schedule on a regular machine with an increasing of the objective function by at most a fixed factor.

In Sect. 8 we will start by presenting an approximation algorithm for the problem $P|r_j|\sum C_j$, see [10]. This algorithm improves the previous best bound given in [8] that is not too efficient since it uses a polynomial approximation scheme for makespan due to Hochbaum and Shmoys [21]. The idea of that result is to divide all the processing times by m, where m is the number of parallel identical machines, and then to use a general conversion algorithm developed for this case. Then we give the explanation of the DELAY-LIST algorithm, that, given an algorithm to find a ρ-approximation for the single machine case, produces a m-machine schedule, with a certain approximation, see [9]. Chekuri et al. gave an algorithm, that uses the DELAY-LIST, with a better guarantee. It is important to note that the DELAY-LIST algorithm gives some schedules which are good both for the average completion time and the makespan, the existence of such schedules was shown also in [8, 43].

Finally in Sect. 9, we present an approximation algorithm for the problem $P|d_{ij}|\sum w_j C_j$, proved in [35]. This problem generalizes problems with precedence constraints and release dates, since they can be represented by d_{ij}. Even special cases of this problem, as $P2||\sum w_j C_j$, are NP-hard, see [7, 25, 27, 28] for other examples. In order to prove this result, we pass through an LP relaxation that is a direct extension of those of Hall et al. in [19] and Schulz in [37], by using stronger inequalities given in [37, 46]. Then a general list scheduling algorithm is used and, listing the jobs by midpoints, it is possible to find an approximation ratio for the general problem with d_{ij}'s.

2 List of Main Results

In this section we summarize, in Table 1, the best-known results for the problems we will study. We just note that in the Sect. 8 we study the problem $P|r_j|\sum C_j$, and we present an algorithm that is a consequence of the DELAY-LIST algorithm, while the best approximation known, for the more general weighted problem, is given by Skutella in [40].

Table 1 List of problems and the current best results

Problem	Best-known approximation	Sections	References
$1|r_j|\sum C_j$	$\frac{e}{e-1} \approx 1.58$	6	[10]
$1|r_j|\sum w_j C_j$	1.6853	5	[16]
$1|r_j, prec|\sum w_j C_j$	$\frac{\sqrt{e}}{\sqrt{e}-1} \approx 2.54$	7	[41]
$P|r_j|\sum w_j C_j$	2	8	[40]
$P|d_{ij}|\sum w_j C_j$	4	9	[35]

3 LP Relaxations for the Release Dates Case

In this section we will present two different relaxed formulations for the easiest problem we will study, namely $1|r_j|\sum w_j C_j$, and their relations with a preemptive schedule that can be constructed in polynomial time. The first one studied will be a time-indexed relaxation formulated by Dyer and Wolsey in [13], while the second one is a mean busy time relaxation, proposed the first time by Schulz, see [37]. Then we will show their equivalence, and we will give a proof of the supermodularity of the solutions, as shown by Goemans in [14].

We first introduce a preemptive schedule that will be used as a basis for many of our next results. It works by scheduling at any point in time the available job with biggest $\frac{w_j}{p_j}$ ratio, where we can assume the jobs to be indexed so that $\frac{w_1}{p_1} \geq \frac{w_2}{p_2} \geq \cdots \geq \frac{w_n}{p_n}$ and ties are broken according to this order. Since when a job j is released and the job currently processed (if any) is k with $j < k$, then we preempt k, this rule yields a preemptive schedule. In the next sections, we will call this schedule LP schedule or fractional schedule.

In [24], Labetoulle et al. proved that the problem $1|r_j, pmtn|\sum w_j C_j$ is (strongly) NP-hard, so this LP schedule does not solve the problem in the preemptive case, but we know, see [38], that the solution is always within a factor of 2 of the optimum and this bound is tight.

Proposition 1.1 *The LP schedule can be constructed in $O(n \log n)$ time.*

Proof Let us create a priority queue of the available jobs not yet completed using as key the ratio $\frac{w_j}{p_j}$ and using a field to express the remaining processing time (for some properties of priority queues see [11]). Then, when a job is released we add it to the priority queue, while, when a job is completed, we remove it from the queue. At any moment the element of the queue with highest priority is processed, if the queue is empty we move to the next release date. Since the number of jobs is n we have at most $O(n)$ operations on the queue, each of them can be implemented in $O(\log n)$ time, hence the algorithm has running time $O(n \log n)$.

3.1 Time-Indexed Relaxation

The first relaxation we study is a time-indexed relaxation. In [13], there were defined two possible time-indexed relaxations, for our aims we just need the weaker one. In this time-indexed relaxation, we define two types of variables: s_j, representing the starting time of job j, and $y_{j,\tau}$, representing the processing of a job j in the interval defined by τ, so that

$$y_{j,\tau} = \begin{cases} 1 & \text{job } j \text{ is processed in } [\tau, \tau + 1) \\ 0 & \text{otherwise} \end{cases}$$

For the sake of simplicity, since we consider nonpreemptive schedule, we replace s_j by C_j just by adding p_j, that gives an equivalent relaxation. Let T be an upper bound of the makespan of an optimal schedule, we can assume $T = \max_j r_j + \sum p_j$. Then we can define the linear program

$$
\begin{aligned}
Z_D = \quad & \min \sum_{j \in N} w_j C_j \\
& \text{subject to} \\
& \sum_{j : r_j \le \tau} y_{j,\tau} \le 1 && \tau = 0, 1, \ldots, T \\
\text{D} \quad & \sum_{\tau = r_j}^{T} y_{j,\tau} = p_j && j \in N \\
& C_j = \frac{p_j}{2} + \frac{1}{p_j} \sum_{\tau = r_j}^{T} \left(\tau + \frac{1}{2}\right) y_{j,\tau} \quad j \in N && (*) \\
& 0 \le y_{j,\tau} && j \in N, \tau = r_j, r_j + 1, \ldots, T.
\end{aligned}
$$

In the LP we have that the equation for C_j corresponds to the correct completion time value of job j when it is nonpredempted, i.e., $y_{j,s_j} = y_{j,s_j+1} = \cdots = y_{j,s_j+p_j-1} = 1$, because

$$
\frac{p_j}{2} + \frac{1}{p_j} \sum_{\tau = r_j}^{T} \left(\tau + \frac{1}{2}\right) y_{j,\tau} = \frac{p_j}{2} + \frac{1}{p_j} \left(s_j + \frac{1}{2} + \cdots + \left(s_j + p_{j-1} + \frac{1}{2} \right) \right) =
$$

$$
= \frac{p_j}{2} + s_j + \frac{1}{p_j} \left(p_j \frac{p_j}{2} \right) = \frac{p_j}{2} + s_j + \frac{p_j}{2} = s_j + p_j = C_j.
$$

The number of variables is pseudo-polynomial. Then, substituting C_j by the equality in constraint $(*)$, we get a transportation problem, see [13], for which we know:

Lemma 1.1 *There is an optimal solution for the LP program D that satisfies* $y_{j,\tau} \in \{0, 1\}$ *for all* j *and* τ.

By the work of Posner, in [31], the time-indexed LP relaxation can be solved in $O(n \log n)$. One can derive a feasible solution y^{LP} to D from the LP schedule by letting $y^{LP}_{j,\tau}$ be equal 1 if job j is being processed in $[\tau, \tau + 1)$, and 0 otherwise.

Theorem 1.1 *The solution* y^{LP} *derived from the LP schedule is an optimal solution for the linear program D.*

Proof We use an exchange argument to prove the theorem. So consider any optimum 0/1 solution y^* of D such that we have $j < k$ and $\sigma > \tau \ge r_j$ with $y^*_{j,\tau} = y^*_{k,\tau} = 1$. By replacing $y^*_{j,\sigma}$ and $y^*_{k,\tau}$ to 0, and $y^*_{k,\sigma}$ and $y^*_{j,\tau}$ to 1, we obtain another feasible solution in which we increase the objective function by $(\sigma - \tau)\left(\frac{w_k}{p_k} - \frac{w_j}{p_j}\right) \le 0$. Hence another optimal solution in which the number of unordered pairs of pieces of jobs is decreased by 1. Repeating this interchange argument we have, at the end, that there exists an optimum solution y' with no $j < k$ and $\sigma > \tau \geq r_j$ such that $y'_{j,\sigma} = y'_{k,\tau} = 1$, hence the solution y' corresponds to the LP schedule.

The linear program D solves neither the problem $1|r_j|\sum w_j C_j$ nor, in general, $1|r_j, pmtn|\sum w_j C_j$, but we can find an optimum solution efficiently despite the big number of variables.

3.2 Mean Busy Time Relaxation

In the mean busy time relaxation, we define an indicator function $I_j(t)$ that takes value 1 if job j is processed at time t and 0 otherwise. In order to exclude possible degenerate cases we assume that, when the machine starts processing a job, it does so for a positive amount of time. Then we can define the mean busy time

$$M_j = \frac{1}{p_j} \int_{r_j}^{T} I_j(t)\, t\, dt$$

to be the average time at which the machine is processing j.

Lemma 1.2 *Let C_j and M_j be, respectively, the completion time and the mean busy time of job j for any nonpreemptive schedule. Then it follows $C_j \geq M_j + \frac{p_j}{2}$. Moreover we have equality if and only if job j is not preempted.*

Proof If job j is processed nonpreemptively if $C_j - p_j \leq t \leq C_j$ we have $I_j(t) = 1$, otherwise 0, hence $C_j = M_j + \frac{p_j}{2}$. Suppose now job j is not processed for some interval in $[C_j - p_j, C_j]$. We know $\int_{r_j}^{T} I_j(t)\, dt = p_j$, so job j is processed for some interval before $C_j - p_j$. Thus

$$M_j = \frac{1}{p_j} \int_{r_j}^{C_j} I_j(t)\, t\, dt < \frac{1}{p_j} \int_{C_j - p_j}^{C_j} t\, dt = \frac{1}{p_j} \left(\frac{C_j^2}{2} - \frac{C_j^2 + p_j^2 - 2\, C_j\, p_j}{2} \right),$$

hence $M_j < C_j - \frac{p_j}{2}$.

In order to give another important property of M_j we have to give some definitions: let $S \subseteq N$ be a set of jobs, then we have $p(S) := \sum_{j \in S} p_j$, $r_{min}(S) := \min_{j \in S} r_j$, and $I_S(t) = \sum_{j \in S} I_j(t)$. Since only one job can be processed at any time we have $I_S(t) \in \{0, 1\}$, so this parameter can be seen as the indicator function for the job set S, hence we can define the mean busy time of the set S as $M_S := \frac{1}{p(S)} \int_0^T I_S(t)\, t\, dt$. From this definition, it follows that the mean busy time of S is just a nonnegative combination of the mean busy times of its elements:

$$p(S) M_S = \int_0^T \left(\sum_{j \in S} I_j(t) \right) t\, dt = \sum_{j \in S} \int_0^T I_j(t)\, t\, dt = \sum_{j \in S} p_j M_j.$$

So we can, now, prove the validity of the shifted parallel inequalities:

Lemma 1.3 *Let S be a set of jobs and consider any nonpreemptive schedule, then*

$$\sum_{j \in S} p_j M_j \geq p(S) \left(r_{min}(S) + \frac{p(S)}{2} \right) \tag{1}$$

holds. Moreover we have equality if the jobs in S are scheduled continuously from $r_{min}(S)$ *to* $r_{min}(S) + p(S)$.

Proof We know that $I_S(t) = 0$ for $t < r_{min}(S)$ and $I_S(t) \leq 1$ for each t, that $\int_{r_{min}(S)}^{T} I_S(t)\, dt = p(S)$, and that $\sum_{j \in S} p_j M_j = p(S) M_S = \int_{r_{min}(S)}^{T} I_S(t)\, t\, dt$. So, from these constraints, it follows that M_S is minimized when $I_S(t) = 1$ for $t \in [r_{min}(S), r_{min}(S) + p(S)]$ and 0 otherwise, that is when jobs in S are processed from $r_{min}(S)$, without interruption. The lower bound of M_S is $r_{min}(S) + \frac{1}{2} p(S)$, so $p(S) \left(r_{min}(S) + \frac{1}{2} p(S) \right)$ minimizes $\sum_{j \in S} p_j M_j$ in every feasible preemptive schedule.

Hence the linear program is

$$Z_R = \min \sum_{j \in N} w_j \left(M_j + \frac{p_j}{2} \right)$$

(R) subject to

$$\sum_{j \in S} p_j M_j \geq p(S) \left(r_{min}(S) + \frac{p(S)}{2} \right) \quad \text{for any } S \subseteq N$$

and it provides a lower bound on the optimum value for $1|r_j, pmtn| \sum w_j C_j$, hence also on the optimum value of the problem $1|r_j| \sum w_j C_j$.

Let us now consider the schedule that, given a set of jobs S, processes these jobs as early as possible. This schedule partitions the set S into $\{S_1, \ldots, S_k\}$, called canonical decomposition of S, and the machine processes the set S in disjoint intervals $[r_{min}(S_l), r_{min}(S_l) + p(S_l)]$ for $l = 1, \ldots, k$. We call a set S *canonical* if it corresponds to its canonical decomposition, that is, all jobs in S are processed in $[r_{min}(S), r_{min}(S) + p(S))$. We can now reformulate Lemma 1.3 by defining, for the canonical decomposition $\{S_1, \ldots, S_k\}$ of $S \subseteq N$, the parameter

$$h(S) := \sum_{l=1}^{k} p(S_l) \left(r_{min}(S_l) + \frac{p(S_l)}{2} \right). \tag{2}$$

Thus, the relaxation R can be written

$$min \left\{ \sum_{j \in N} w_j \left(M_j + \frac{p_j}{2} \right) : \sum_{j \in S} p_j M_j \geq h(S) \text{ for all } S \subseteq N \right\}. \tag{3}$$

Theorem 1.2 *Let M_j^{LP} be the mean busy time of job j in the LP schedule, then M^{LP} is an optimum solution of the linear program R.*

Proof By Lemma 1.3 we know M^{LP} is a feasible solution for the linear program R. Recall the jobs are indexed by nonincreasing $\frac{w_j}{p_j}$ ratio. Then, for the set $[i] := \{1, 2, \ldots, i\}$, let $S_1^i, \ldots, S_{k(i)}^i$ denote its canonical decomposition. Assuming $\frac{w_{n+1}}{p_{n+1}} = 0$, then for any vector $M = (M_j)_{j \in N}$ we have

$$\sum_{j \in N} w_j M_j = \sum_{i=1}^{n} \left(\frac{w_i}{p_i} - \frac{w_{i+1}}{p_{i+1}} \right) \sum_{j \leq i} p_j M_j = \sum_{i=1}^{n} \left(\frac{w_i}{p_i} - \frac{w_{i+1}}{p_{i+1}} \right) \sum_{l=1}^{k(i)} \sum_{j \in S_l^i} p_j M_j. \tag{4}$$

Hence we expressed $\sum w_j M_j$ as a nonnegative combination of summands over canonical sets. By construction the LP schedule continuously processes the jobs in the canonical sets S_l^i in the interval $\left[r_{min} \left(S_l^i \right), r_{min} \left(S_l^i \right) + p \left(S_l^i \right) \right]$. So for any canonical set and each feasible solution M of the linear program R it follows, by Lemma 1.3,

$$\sum_{j \in S_l^i} p_j M_j \geq h \left(S_l^i \right) = p \left(S_l^i \right) \left(r_{min} \left(S_l^i \right) + \frac{p \left(S_l^i \right)}{2} \right) = \sum_{j \in S_l^i} p_j M_j^{LP}. \tag{5}$$

So together with Eq. (4), we derive a lower bound on $\sum p_j M_j$ for each feasible solution M of R, that it is tight for the LP schedule.

We, now, prove the equivalence of the two relaxations:

Corollary 1.1 *For any weight $w \geq 0$ we have that the two LP relaxations D and R give the same optimal value, that is $Z_D = Z_R$.*

Proof Given the solution y^{LP} to R we can express the mean busy time M_j^{LP} of any job as

$$M_j^{LP} = \frac{1}{p_j} \sum_{\tau = r_j}^{T} y_{j,\tau}^{LP} \left(\tau + \frac{1}{2} \right).$$

So by Theorems 1.1 and 1.2 we have the desired equality.

We recall, before studying some polyhedral consequences of the LP formulations, that the LP schedule solves optimally $1|r_j| \sum w_j M_j$ over the preemptive schedules but it does not, generally, minimize $1|r_j| \sum w_j C_j$ over the preemptive schedules.

3.3 Polyhedral Consequences

We study now some polyhedral consequences of the two LP relaxations. Let P_D^∞ be the feasibility region for D when $T = \infty$. So

$$P_D^\infty := \left\{ y \geq 0 : \sum_{j:r_j \leq \tau} y_{j,\tau} \leq 1 \text{ for } \tau \in \mathbb{N}, \sum y_{j,\tau} = p_j \text{ for all } j \in N \right\}.$$

Similarly the LP relaxation R, with its constraints, defines the polyhedron P_R.

Theorem 1.3 *The polyhedron P_R is the convex hull of the mean busy time vectors M of all the preemptive schedules. Moreover each vertex of P_R is the mean busy time vector of an LP schedule.*

Proof We already proved in Lemma 1.3 that P_R contains the convex hull of the mean busy time vectors M of all feasible preemptive schedules. In order to prove the other direction, we need to show that every extreme point of P_R corresponds to a preemptive schedule and each extreme ray of P_R is a recession direction for the convex hull of mean busy time vectors. Since every extreme point of P_R is the unique minimizer of $\sum w_j M_j$ for some $w \geq 0$ and by Theorem 1.2, it follows that each extreme point of P_R corresponds to a preemptive schedule. In order to finish the proof, we note that the extreme rays of P_R are the n unit vectors of $\mathbb{R}^N$. An extension of the results in [4] to preemptive schedules and mean busy times implies that the directions of recession for the convex hull of mean busy time vectors are those unit vectors of $\mathbb{R}^N$.

Theorem 1.4 *Let be given the mapping $\mathcal{M} : y \mapsto \mathcal{M}(y) \in \mathbb{R}^N$ defined by $\mathcal{M}(y)_j = \frac{1}{p_j} \sum_{\tau \geq r_j} y_{j,\tau} \left(\tau + \frac{1}{2}\right)$ for all $j \in N$. Then we have that P_R is the image of P_D^∞ in the space of M-variables under $\mathcal{M}$.*

Proof Let $y \in P_D^\infty$ and $S \subseteq N$ with canonical decomposition $\{S_1, \ldots, S_k\}$. Then, by definition of $\mathcal{M}(y)_j$ it follows:

$$\sum_{j \in S} p_j \mathcal{M}(y)_j = \sum_{j \in S} \sum_{\tau \geq r_j} y_{j,\tau} \left(\tau + \frac{1}{2}\right) \geq \sum_{l=1}^{k} \sum_{\tau = r_{min}(S_l)}^{r_{min}(S_l)+p(S_l)} \left(\tau + \frac{1}{2}\right)$$
$$= \sum_{l=1}^{k} p(S_l) \left(r_{min}(S_l) + \frac{p(S_l)}{2}\right) = h(S),$$

where the inequality follows from the definition of P_D^∞ and by applying an interchange argument as we already did in Theorem 1.1. Hence $\mathcal{M}(y) \in P_R$ and $\mathcal{M}(P_D^\infty) \subseteq P_R$. In order to prove the other direction, we recall that P_R can be represented as the sum of the convex hull of all mean busy time vectors of LP schedules and the nonnegative orthant. We know that the mean busy time vector M^{LP} of any LP schedule is the projection of the corresponding vector y^{LP}, hence we only need to prove that every unit vector e_j is a recession direction for $\mathcal{M}(P_D^{\infty})$. In order to prove it, we can fix an LP schedule where we denote the associated $0/1$ vector and the mean busy time vector, respectively, by y^{LP} and $M^{LP} = \mathcal{M}(y^{LP})$. Then, for any job $j \in N$ and any real $\lambda > 0$, we have to prove $M^{LP} + \lambda e_j \in \mathcal{M}(P_D^{\infty})$. Let now $\tau_{max} = \max\{\tau : y_{k,\tau}^{LP} = 1 : k \in N\}$, then we choose θ so that $y_{j,\theta}^{LP} = 1$ and $\mu > \max\{\lambda p_j, \tau_{max} - \theta\}$. Hence it is possible to define $y'_{j,\theta} = 0$, $y'_{j,\theta+\mu} = 1$; and $y'_{k,\tau} = y_{k,\tau}^{LP}$ otherwise. In the associated preemptive schedule we postpone the processing of job j by μ units, so from interval $[\theta, \theta + 1)$ to $[\theta + \mu, \theta + \mu + 1)$, hence $M' = \mathcal{M}(y')$ equals $M'_j = M_j^{LP} + \frac{\mu}{p_j}$ and $M'_k = M_k^{LP}$ for $k \neq j$. Calling $\lambda' = \frac{\mu}{p_j} \geq \lambda$, we have $M' = M^{LP} + \lambda' e_j$, hence $M^{LP} + \lambda' e_j$ is a convex combination of M^{LP} and M'. Let now y correspond to a convex combination of y^{LP} and y'. We conclude saying that $M^{LP} + \lambda e_j = \mathcal{M}(y) \in \mathcal{M}(P_D^{LP})$ by the convexity of P_D^{LP}, so the implication $y \in P_D^{LP}$ follows.

As last result about the LP relaxations, we note that the feasible set P_R defines a particular polyhedron.

Proposition 1.2 *Given the function h as defined in Eq.* (2), *it defines a supermodular set function.*

Proof Let $j, k \in N$ be two jobs and consider any subset $S \subseteq N \setminus \{j, k\}$. By using a job-based method, hence considering job k after jobs in S, we construct an LP schedule minimizing $\sum_{i \in S \cup \{k\}} p_i M_i$. We know, from the definition of h, that considering the mean busy time M^{LP} we have $\sum_{i \in S} p_i M_i^{LP} = h(S)$ and $\sum_{i \in S \cup \{k\}} p_i \cdot M_i^{LP} = h(S \cup \{k\})$. By construction we have that job k is scheduled in the first p_k units of idle time after r_k, hence we can see M_k^{LP} as the mean of these time units. The next step in the proof is to construct another LP schedule, with mean busy time vector $\widetilde{M}^{LP}$, considering before set S, then job j and at the end job k, in such a way to have $\widetilde{M}_i^{LP} = M_i^{LP}$ for $i \in S$, $\sum_{i \in S \cup \{j\}} p_i \widetilde{M}_i^{LP} = h(S \cup \{j\})$ and $\sum_{i \in S \cup \{j,k\}} p_i \widetilde{M}_i^{LP} = h(S \cup \{j, k\})$. For sure job k cannot have processing time earlier than the former LP schedule, since there only S was considered, and now $S \cup \{j\}$, hence we have $\widetilde{M}_k^{LP} \geq M_k^{LP}$ that implies the supermodularity, because

$$h(S \cup \{j, k\}) - h(S \cup \{j\}) = \widetilde{M}_k^{LP} \geq M_k^{LP} = h(S \cup \{k\}) - h(S).$$

From Proposition 1.2 it follows that the job-based method for the construction of the LP schedule is like the greedy algorithm that minimizes $\sum w_j M_j$ over the supermodular P_R.

4 Conversion Algorithm

In this section we will explain a general algorithm that converts the preemptive LP schedule into a nonpreemptive one, by using the α-points of the jobs, introduced for the first time by Phillips et al. in [30]. Then we will show an example in such a way to understand better how the algorithm works. After we give some easy bounds that will be improved in the next sections by Goemans et al. see [16]. Finally, we will also present a general technique to derandomize the algorithm, as shown by Goemans in [15].

4.1 The Algorithm

We start by defining what is the α-point of a job j, denoted by $t_j(\alpha)$. Given $0 < \alpha \leq 1$ the *α-point* of job j is the first point in time when job j has been processed for a total of $\alpha \cdot p_j$ time units, so an α-fraction of job j is completed. Trivially we can denote the starting and the completion time of job j as, respectively, $t_j(0^+)$ and $t_j(1)$.

By definition, the mean busy time M_j^{LP} of job j in the LP schedule is the average value of all its α-points, so

$$M_j^{LP} = \int_0^1 t_j(\alpha)\, d\alpha. \tag{6}$$

We also define, for a given α, $0 < \alpha \leq 1$, and a fixed job j, the parameter $\eta_k(j)$ to be the fraction of job k that is completed in the LP schedule by $t_j(\alpha)$. It can be seen easily that $\eta_j(j) = \alpha$, since at $t_j(\alpha)$ the processed fraction of job j is α.

By the construction of the preemptive schedule, there is no idle time between the start and the completion of any job j, so, if there is some idle time, it must be before $t_j(0^+)$. This quantity depends on the specific job, hence we can denote by τ_j the idle time that occurs between time 0 and the starting time of job j in the LP schedule. By the previous observations, we can write the α-point of j as

$$t_j(\alpha) = \tau_j + \sum_{k \in N} \eta_k(j) \cdot p_k. \tag{7}$$

For a given α, $0 < \alpha \leq 1$, consider the α-schedule which processes the jobs as early as possible in order of nondecreasing α-point and in a nonpreemptive way. Then the completion time of job j in this α-schedule is denoted by C_j^α. The conversion algorithm, called α-CONV, works as follows:

> Consider the jobs $j \in N$ in order of nonincreasing α-points and iteratively change the preemptive LP schedule to a nonpreemptive schedule by applying the following steps:
>
> - remove the $\alpha \cdot p_j$ units of job j processed before $t_j(\alpha)$ from the machine and make them idle time, we say this idle time is *caused by* job j;
> - delay the whole processing done after the α-point of job j by p_j units of time;

- remove the remaining units of job j from the machine (they are $(1-\alpha)\cdot p_j$) and move earlier the processings occurring later. Hence job j is processed in a nonpreemptive way exactly in the interval $[t_j(\alpha), t_j(\alpha)+p_j]$.

The feasibility of this schedule follows by the fact that the jobs are scheduled in nondecreasing order of α-point, so no job is started before $t_j(\alpha) \geq r_j$.

4.2 Example

We provide an example of the conversion algorithm. Let $\alpha = \frac{1}{2}$ and let $N = \{j_1, j_2, j_3, j_4\}$ be the set of jobs, with $\frac{w_1}{p_1} \leq \cdots \leq \frac{w_4}{p_4}$, so that:

- Job 1: has $p_1 = 5$ and $r_1 = 0$, then we obtain $t_1(\alpha) = t_1 = 2.5$ and $C_1 = 10$;
- Job 2: has $p_2 = 1$ and $r_2 = 3$, then we obtain $t_2(\alpha) = t_2 = 3.5$ and $C_2 = 4$;
- Job 3: has $p_3 = 3$ and $r_3 = 5$, then we obtain $t_3(\alpha) = t_3 = 6.5$ and $C_3 = 9$;
- Job 4: has $p_4 = 1$ and $r_4 = 7$, then we obtain $t_4(\alpha) = t_4 = 7.5$ and $C_4 = 8$.

In Fig. 1, it is shown the LP schedule, with the relatives α-points. Note that, because of the priorities, each newly released job preempts the current one.

In Fig. 2, it is possible to see the individual iterations of the α-CONV algorithm, where at the top we have the LP schedule.

4.3 Some Consequenques

Directly from the algorithm it is easy to find some bound on the completion time of a job. These bounds will be used in the next chapters to obtain improved approximation guarantees.

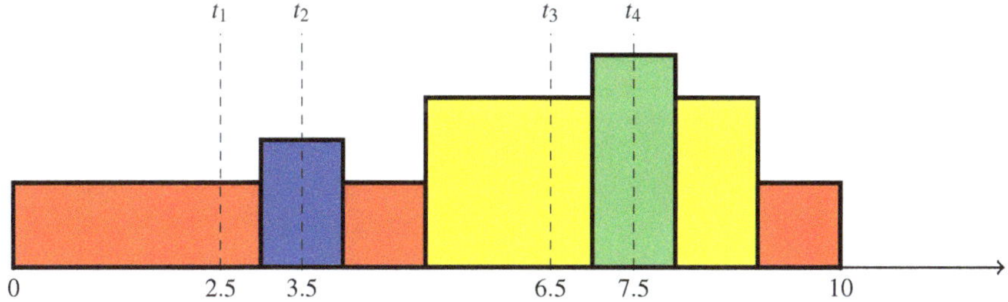

Fig. 1 LP schedule of the example model

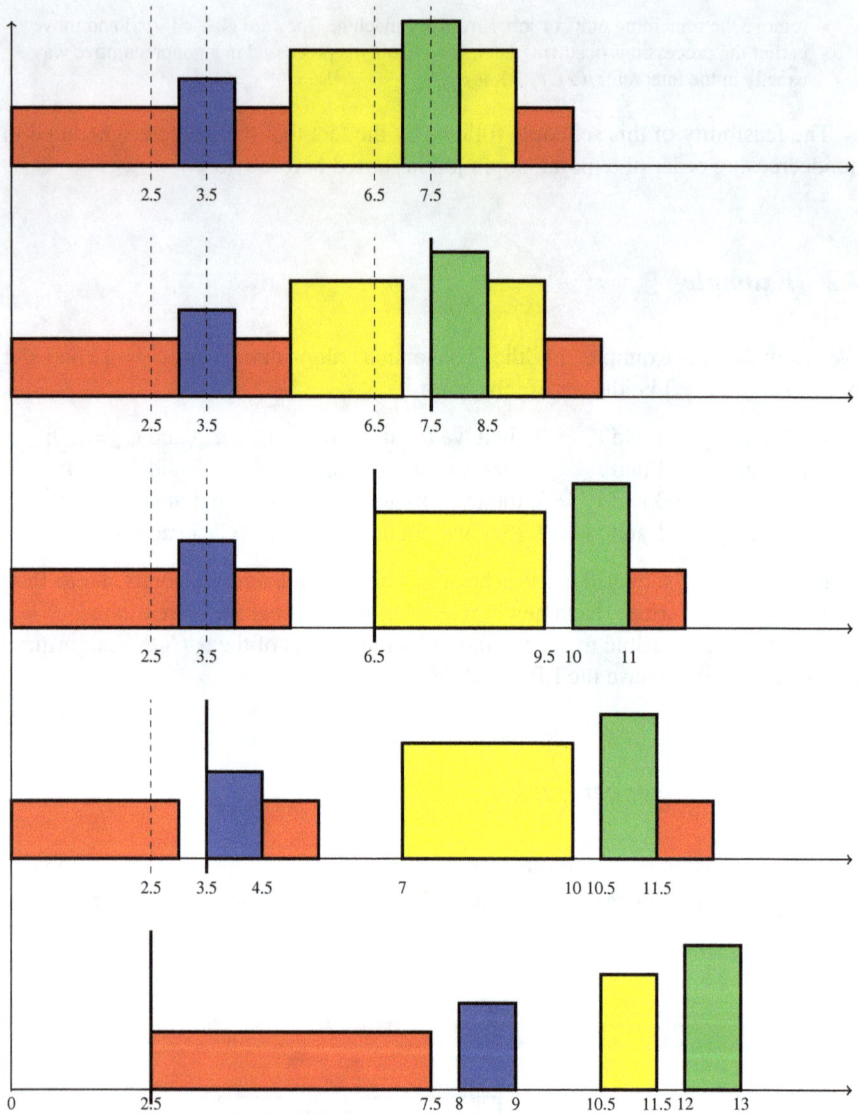

Fig. 2 Iterations of the algorithm

Lemma 1.4 *The algorithm α-CONV set the completion time for job j to be*

$$C_j^\alpha = t_j(\alpha) + \sum_{k:\eta_k(j)\geq\alpha} (1 - \eta_k(j) + \alpha) \cdot p_k. \tag{8}$$

Proof Consider job j, after the algorithm its completion time equals the sum of the processing times of jobs scheduled no later than j and the idle time before $t_j(\alpha)$. By the α-point ordering, we have that, before the completion time of job j, the machine is busy a total of $\sum_{k:\eta_k(j)\geq\alpha} p_k$ time. We now have to reformulate the idle time, since now it is the sum of τ_j and the delay in the start of the other jobs in the conversion algorithm. Each job k such that $\eta_k(j) \geq \alpha$ contributes $\alpha \cdot p_k$ units of idle time, while the other jobs just $\eta_k(j) \cdot \alpha$ units of idle time. Hence the total idle time before the start of job j in α-CONV is

$$\tau_j + \sum_{k:\eta_k(j)\geq\alpha} \alpha \cdot p_k + \sum_{k:\eta_k(j)<\alpha} \eta_k(j) \cdot p_k. \tag{9}$$

Then the completion time of job j in the schedule constructed by the algorithm is the sum of the processing times and the idle time before the completion of job j, thus

$$\begin{aligned} C_j^\alpha &= \sum_{k:\eta_k(j)\geq\alpha} p_k + \tau_j + \sum_{k:\eta_k(j)\geq\alpha} \alpha \cdot p_k + \sum_{k:\eta_k(j)<\alpha} \eta_k(j) \cdot p_k = \\ &= t_j(\alpha) + \sum_{k:\eta_k(j)\geq\alpha} (1 - \eta_k(j) + \alpha) \cdot p_k, \end{aligned} \tag{10}$$

where the second equality follows from (7).

Corollary 1.2 *In an α-schedule, we have the following bound for the completion time of job j:*

$$C_j^\alpha \leq t_j(\alpha) + \sum_{k:\eta_k(j)\geq\alpha} (1 - \eta_k(j) + \alpha) \cdot p_k. \tag{11}$$

Proof Since the schedule constructed by α-CONV processes the jobs in order of nondecreasing α-point and by Lemma 1.4, the result follows easily.

We will see in the next section that for a fixed α, $0 < \alpha \leq 1$, the value of the α-schedule is within a factor of $\max\{1 + \frac{1}{\alpha}, 1 + 2\alpha\}$ of the optimum LP value, so it is minimized for $\alpha = \frac{1}{\sqrt{2}}$ and the bound is $1 + \sqrt{2}$. We will see how to beat this bound by choosing α randomly from some probability distribution.

We explain here how is it possible to derandomize easily the algorithm. We prove it now since it is a general idea and it can be used for all the LP-based schedules with random $\alpha \in (0, 1]$.

Lemma 1.5 *Over all choices of α, there are at most n different orderings given by the algorithm.*

Proof We have changes in the α-schedule only when a job is preempted. Thus the total number of changes in the α-schedule is bounded from above by the total number of preemptions. From the fact that a preemption in the LP can occur only when a new job is released, we have that the total number of preemptions is at most $n - 1$, so there are at most n different orderings.

Proposition 1.3 *There are at most n different schedules and they can be computed in $O(n^2)$ time.*

Proof By Lemma 1.5, we have that the total number of preemptions is at most $n-1$, hence there are at most n different orderings, i.e., combinatorial values of α. Since we can compute each α-schedule in $O(n)$ time the total running time is at most $O(n^2)$.

Obviously the schedule can be constructed by using job-dependent α_j, but we preferred to explain for a common α in such a way to keep the notation easier, anyway the generalization to the job-dependent case is straightforward.

5 Approximations for $1|r_j|\sum w_j C_j$

In this section, we will see some approximation algorithms for the scheduling problem $1|r_j|\sum w_j C_j$. Starting from the preemptive LP optimal solution, we will use a nonpreemptive feasible schedule, so we can relate the value to $Z_D = Z_R$, in such a way to obtain some specific approximation guarantee. In Sect. 5.1 we will prove the results given by Goemans in [15], namely the best possible guarantee of $1+\sqrt{2}$ for fixed α and a 2-approximation algorithm by choosing α randomly. We will present also an easy way to derandomize this algorithm. In Sect. 5.2, following the ideas of Goemans et al. in [16], we will derive another proof for the 2-approximation performance, but using job-dependent α_j values. In the same paper the authors proved stronger approximation algorithms, involving some specific probability distribution; we will present them in Sects. 5.3 and 5.4. These algorithms are, respectively, for a common α and for job-dependent α_j, and we will prove that, with respect to those probability distributions, they are the best possible. Finally in Sect. 5.5 we give an example showing that the conversion algorithm based on the LP relaxations used can reach an approximation guarantee of at most $\frac{e}{e-1} \approx 1.5819$, see [16].

5.1 2-Approximation Algorithm

Given the LP schedule and $\alpha \in (0, 1]$, the algorithm A_α orders the jobs in order of increasing α-points and it schedules the jobs as early as possible following this order. For a given α, $0 < \alpha \le 1$, a valid α-schedule is the schedule which processes the jobs in order of increasing $t_j(\alpha)$ and so that job j is not processed before $t_j(\alpha)$, so an α-schedule is valid if it is the output of a conversion algorithm.

For any set of jobs $S \subseteq N$ let us recall that the processing time of S is the sum of processing times of the jobs in S, i.e., $p(S) = \sum_{j \in S} p_j$.

In order to simplify the notation, since we use a common α, we may denote the α-point of job j just by t_j.

Proposition 1.4 *If the preemptive schedule processes all the jobs in S in the interval* $[0, p(S)]$*, then there exists a valid* α*-schedule which processes the jobs in S in the interval* $[\alpha p(S), (\alpha+1)p(S)]$.

Proof Given the set S, with $|S| = s$, we can reindex the jobs in such a way that $t_1 \le t_2 \le \cdots \le t_s$. For any index i, at time t_i, the preemptive schedule can have processed at most all the jobs $1, 2, \ldots, i-1$ completely and the jobs $i+1, \ldots, s$ for an α-fraction. So it follows that

$$t_i(\alpha) \le p_1 + \cdots + p_{i-1} + \alpha p_i + \cdots + \alpha p_s = \sum_{k=1}^{i-1} p_k + \alpha \sum_{k=i}^{s} p_k \le \sum_{k=1}^{i-1} p_k + \alpha p(S)$$

Hence job j_1 can start at time $\alpha p(S)$, job j_2 at time $p_1 + \alpha p(S)$, and so on. So the jobs can be scheduled consecutively in this order in the interval $[\alpha p(S), (\alpha+1)p(S)]$, and it is a valid α-schedule.

This proposition was specific for the case if the set S started to be processed at time 0. For the general case we have to define before some values: given a canonical set $S \subseteq N$, let $s(S)$ be the first time when jobs in S are processed and let $\mu_S(j)$ be the fraction of job j that is processed in the preemptive schedule before $s(S)$. Then we have following statement.

Proposition 1.5 *Given any canonical set S, there is a valid* α*-schedule which processes the jobs in S in the interval* $[t(S), t(S) + p(S)]$*, where we have to define* $t(S) := \max\{(1+\frac{1}{\alpha})s(S), s(S) + \alpha p(S)\}$.

In order to have the desired approximation guarantee, we prove a stronger statement.

Proposition 1.6 *Let S be any canonical set, all the jobs in S are processed, by a valid* α*-schedule, in the interval* $[t(S), t(S) + p(S)]$*, where, in this case, we have* $t(S) := s(S) + \max\{\sum_{j:\mu_S(j)\ge\alpha} p_j, \alpha p(S)\}$.

Proof From the definition of the algorithm A_α, before to start the process of a job in S it is required exactly $\sum_{j:\mu_S(j)\ge\alpha} p_j$ time, so the sum of the processing times of the jobs with $t_j \le t_k$, where k is the first job processed in S. We can see, that for all those jobs $t_j \le s(S)$, otherwise, since S is a canonical set the jobs j would be scheduled before k. So these jobs can be scheduled in the interval $[s(S), s(S) + \sum_{j:t_j\le t_k} p_j]$. If $s(S) + \sum_{j:t_j\le t_k} p_j \ge s(S) + \alpha p(S)$, then we have, according to a shifted version of Proposition 1.4, that jobs in S can follow immediately. In the other case, if $s(S) + \sum_{j:t_j\le t_k} p_j < s(S) + \alpha p(S)$, it is just needed to delay the start of the jobs preceding S by $\alpha p(S) - \sum_{j:t_j\le t_k} p_j$ in order to have the desired result.

Observation Proposition 1.6 implies Proposition 1.5.

Proof We know, by definition, that the μ_S fraction of every job is processed before than $s(S)$; that means $\sum \mu_S(j)p_j \leq s(S)$. Thus it follows that $\sum_{j:\mu_S(j)\geq\alpha} \cdot p_j \leq \sum \frac{\mu_S(j)}{\alpha} p_j \leq \frac{s(S)}{\alpha}$.

From these propositions, the two main results follow easily.

Theorem 1.5 *The algorithm A_α is a ρ-approximation for $1|r_j|\sum w_j C_j$, where $\rho = \max\{1+\frac{1}{\alpha}, 1+2\alpha\}$. In particular for $\alpha = \frac{1}{\sqrt{2}}$ the performance guarantee is $1+\sqrt{2}$.*

Proof For any canonical set S, according to Proposition 1.5, we have that the mean busy time given by A_α satisfies

$$\begin{aligned}\tilde{M}(S) &\leq \max\{(1+\frac{1}{\alpha})s(S)\ ,\ s(S)+\alpha p(S)\} + \frac{p(S)}{2}\\ &\leq (1+\frac{1}{\alpha})s(S) + (\alpha+\frac{1}{2})p(S) \leq \max\{1+\frac{1}{\alpha}\ ,\ 1+2\alpha\}\left(s(S)+\frac{p(S)}{2}\right)\end{aligned}$$

and the result follows from the mean busy time LP relaxation defined in Sect. 3.2. It is easily seen that $\max\{1+\frac{1}{\alpha}\ ,\ 1+2\alpha\}$ is minimized for $\alpha = \frac{1}{\sqrt{2}}$. Hence $A_{1/\sqrt{2}}$ is a $1+\sqrt{2}$-approximation algorithm.

We can modify the algorithm A_α by choosing a random value of α, so the algorithm R_α behaves in the same way of A_α with the difference that α is chosen accordingly to a uniform distribution in $(0, 1]$.

Theorem 1.6 *The algorithm R_α outputs a schedule with expected weight at most $2 \cdot OPT$.*

Proof For any canonical set S, its expected mean busy time is

$$\begin{aligned}E\left[\tilde{M}(S)\right] &\leq E\left[\max\{s(S)+\sum_{j:\mu_S(j)\geq\alpha} p_j\ ,\ s(S)+\alpha p(S)\} + \frac{p(S)}{2}\right] \leq\\ &\leq s(S)+\frac{p(S)}{2} + E\left[\sum_{j:\mu_S(j)\geq\alpha} p_j\right] + E\left[\alpha p(S)\right] =\\ &= s(S)+\frac{p(S)}{2} + \sum Pr\left[\alpha \leq \mu_S(j)\right]\cdot p_j + E[\alpha]\cdot p(S) =\\ &= s(S)+p(S)+\sum \mu_S(j)p_j \leq 2s(S)+p(S) = 2\left(s(S)+\frac{p(S)}{2}\right)\end{aligned}$$

by definition of $\mu_S(j)$ and Proposition 1.6.

In order to derandomize the algorithm, we have two different ways: the first, that is general, is explained in Sect. 4, by Lemma 1.5 and Proposition 1.3, while the second applies for our case, it returns a slightly smaller guarantee, but, given a fixed integer k, it runs in $O(n\log n + kn)$ time, so it is faster.

Proposition 1.7 *There is a deterministic algorithm with a guarantee of* $\left(2+\frac{1}{k}\right)$ *running in* $O(n\log n + kn)$ *time.*

Proof Let k be a fixed integer, then for $\alpha_i = \frac{i}{k+1}$, for $i = 1, \dots, k$, we run the algorithm A_{α_i}. The best schedule will be at least as good as the schedule returned by the randomized algorithm which selects the value of α uniformly between the k values α_i. Following the proof of Theorem 1.6 we derive that, for this randomized algorithm, we have:

$$\begin{aligned} E[\tilde{M}(S)] &\le s(S) + \frac{p(S)}{2} + \sum \Pr(\alpha \le \alpha_j)\cdot p_j + E[\alpha p(S)] \\ &\le s(S) + p(S) + \frac{k+1}{k}\sum \alpha_j p_j \le \left(2+\frac{1}{k}\right)s(S) + p(S) \\ &\le \left(2+\frac{1}{k}\right)\left(s(S)+\frac{p(S)}{2}\right). \end{aligned}$$

that implies the performance guarantee.

The running time is just $O(n\log n + kn)$ since $O(n\log n)$ follows by the algorithm and $O(kn)$ from the fact that we consider only k schedules.

5.2 Another 2-Approximation Algorithm

Here we will derive, by using some probability distribution, a different proof for the approximation guarantee of Sect. 5.1, i.e., a 2-approximation, by using job-dependent α_j values. Then we will refine the analysis in order to have some stricter bounds that will be used after.

Theorem 1.7 *Let* α_j *be random variables drawn from* $(0, 1]$ *pairwise independently and uniformly. Then we have that the expected value of the resulting* (α_j)*-schedule is within a factor of 2 of the optimum LP value. The same result holds for the case of common random* α *drawn uniformly from* $(0, 1]$.

Proof Let us define $E_U[F(\boldsymbol{\alpha})]$ as the expectation of the function F in the variable $\boldsymbol{\alpha}$ when $\boldsymbol{\alpha}$ is uniformly distributed. We prove the statement by using some job-by-job bound and by the linearity of expectation.

Let us consider an arbitrary fixed job j. Suppose, for now, that α_j is fixed and we study $E_U[C_l^{\boldsymbol{\alpha}}|\alpha_j]$. By the fact that α_j and α_k are independent for each $j \ne k$, we have, by Corollary 1.2, the following:

$$\begin{aligned} E_U[C_l^\alpha|\alpha_j] &\le t_j(\alpha_j) + \sum_{k\neq j} p_k \int_0^{\eta_k(j)} (1-\eta_k(j)+\alpha_k)\, d\alpha_k + p_j \\ &= t_j(\alpha_j) + \sum_{k\neq j}\left(\eta_k(j) - \frac{\eta_k(j)^2}{2}\right)\cdot p_k + p_j \\ &\le t_j(\alpha_j) + \sum_{k\neq j} \eta_k(j)\cdot p_k + p_j \le 2\cdot\left(t_j(\alpha_j) + \frac{p_j}{2}\right). \end{aligned}$$

Now we just have to integrate over all possible choices of α_j yielding:

$$E_U[C_j^\alpha] = \int_0^1 E_U[C_j^\alpha|\alpha_j]\, d\alpha_j \le 2\cdot\left(\int_0^1 t_j(\alpha_j)\, d\alpha_j + \frac{p_j}{2}\right) = 2\cdot\left(M_j^{LP} + \frac{p_j}{2}\right).$$

The optimum value, Z_D, is given by $\sum w_j\left(M_j^{LP} + \frac{p_j}{2}\right)$, so we have, by linearity of the expectations, that $E_U[\sum w_j C_j^\alpha] \le 2\cdot Z_D$.

We can start by observing that if a job j is preempted in the LP schedule at time s and it is resumed at time $t > s$ in the interval $[s, t]$, the jobs processed have greater $\frac{w_k}{p_k}$ ratio, so these jobs are completely processed in this interval. Hence, in the LP schedule, in $[s, t]$, the machine is constantly busy, alternating between complete processing of jobs k with $\frac{w_k}{p_k} > \frac{w_j}{p_j}$ and portions of job j. At the same time we can say that any job preempted at the starting time $t_j(0^+)$ of job j must wait, at least until $t_j(1)$, to be resumed. Hence, for any fixed job j, we partition the jobs of $N \setminus \{j\}$ into two subsets: N_2, the set of jobs processed between $t_j(0^+)$ and $t_j(1)$, and N_1, the set of all remaining jobs. We recall two parameters defined before: $\eta_k(j)$ is the fraction of job k completed by time $t_j(\alpha_j)$ and $\mu_j(k)$ is the fraction of job j completed by time the starting time of job k, i.e., by $t_k(0^+)$. If j is fixed and $k \in N_1$, then the function $\eta_k(j)$ is constant, so we define $\eta_k := \eta_k(j)$, otherwise we have

$$\eta_k(j) = \begin{cases} 0 & \text{if } \alpha_j \le \mu_k(j) \\ 1 & \text{if } \alpha_j > \mu_k(j) \end{cases} \quad \text{for } k \in N_2. \tag{12}$$

We can now rewrite the expression (7) for the α_j-point as

$$t_j(\alpha_j) = \tau_j + \sum_{k\in N_1} \eta_k\cdot p_k + \sum_{k\in N_2:\alpha_j>\mu_k(j)} p_k + \alpha_j\cdot p_j. \tag{13}$$

If we insert this reformulation in the formula (6) in order to calculate the LP midpoint it yields

$$M_j^{LP} = t_j(0^+) + \sum_{k\in N_2}(1-\mu_k(j))\cdot p_k + \frac{p_j}{2}; \tag{14}$$

at the same time we can rewrite Corollary 1.2 as

$$C_j^\alpha \le t_j(0^+) + \sum_{k\in N_1:\alpha_k\le\eta_k(j)} (1-\eta_k(j)+\alpha_k)\cdot p_k + \\ + \sum_{k\in N_2:\alpha_j>\mu_k(j)} (1+\alpha_k)\cdot p_k + (1+\alpha_j)\cdot p_j. \tag{15}$$

Here when we considered $k \in N_2$, we used the equivalence between the inequalities $\alpha_k \le \eta_k(j)$ and $\alpha_j > \mu_k(j)$. It is easy to see that, in Eq. (15), small values of α keep the terms of the form $(1-\eta_k+\alpha_k)$ and $(1+\alpha_k)$ small, while increasing the α values decrease the number of summands, so we have to balance the contributes of the two effects to reduce the bound on the expected value of the completion time $E[C_j^\alpha]$.

5.3 Bounds for Common α Value

In this section we will prove some better bounds for the common α case, by using a suitable probability distribution, with a truncated exponential function as density function. Then we will give the better approximation guarantee for this case.

Let $\gamma \approx 0.467$ be the unique solution of

$$1-\frac{\gamma^2}{1+\gamma} = \gamma + \ln(1+\gamma) \tag{16}$$

in the interval (0, 1). We define $c := \frac{1+\gamma}{1+\gamma-e^{-\gamma}} < 1.745$ and we let α be chosen according to the density function:

$$f(\alpha) = \begin{cases} (c-1)\cdot e^\alpha & \text{if } \alpha \le \delta \\ 0 & \text{otherwise.} \end{cases} \tag{17}$$

Since it must be a density function, the integral must be 1, hence we can find the value $\delta := \delta(c) = \ln\left(\frac{c}{c-1}\right)$, so that

$$\int_0^\delta f(\alpha)\,d\alpha = \int_0^\delta (c-1)\cdot e^\alpha\,d\alpha = (c-1)(e^\delta-1) = (c-1)\left(\frac{c}{c-1}-1\right) = 1.$$

We prove now a technical lemma that is the key for the proof of the main theorem in this section. The function f has two important properties bounding the delay of job j; inequality (18) refers to jobs in N_1, while inequality (19) is for the jobs in N_2.

Lemma 1.6 *The above defined function f is a density function satisfying the following properties:*

$$\int_0^\eta f(\alpha) \cdot (1 + \alpha - \eta)\, d\alpha \leq (c - 1) \cdot \eta \quad \textit{for all } \eta \in [0, 1], \tag{18}$$

$$\int_\mu^1 f(\alpha) \cdot (1 + \alpha)\, d\alpha \leq c \cdot (1 - \mu) \qquad \textit{for all } \mu \in [0, 1]. \tag{19}$$

Proof In order to prove the inequality (18) we have to split the interval of possible η into $\eta \in [0, \delta]$, so we get

$$\int_0^\eta f(\alpha)(1 + \alpha - \eta)\, d\alpha = (c - 1) \int_0^\eta e^\alpha (1 + \alpha - \eta)\, d\alpha =$$
$$= (c - 1) \cdot [(e^\eta - 1)(1 - \eta) + \eta e^\eta - e^\eta + 1] = (c - 1) \cdot \eta$$

and $\eta \in (\delta, 1]$, for which we have

$$\int_0^\eta f(\alpha)(1 + \alpha - \eta)\, d\alpha < \int_0^\delta f(\alpha)(1 + \alpha - \eta)\, d\alpha = (c - 1) \cdot \delta < (c - 1) \cdot \eta.$$

For Eq. (19) we have to study $\mu \in (\delta, 1]$, that has integral 0, and $\mu \in [0, \delta]$ for which

$$\int_\mu^1 f(\alpha)(1 + \alpha)\, d\alpha = (c - 1) \int_\mu^\delta e^\alpha (1 + \alpha)\, d\alpha \leq (c - 1) \cdot (\delta e^\delta - \eta e^\eta)$$

here it is easier to pass through a different variable γ, so that we can define $c := \frac{1+\gamma}{1+\gamma-e^{-\gamma}}$, with $\delta = 1 - \frac{\gamma^2}{1+\gamma}$ and $\frac{c-1}{c} = \frac{e^{-\gamma}}{1+\gamma}$. The motivation will be seen later. Now, by substitution, we have

$$\int_\mu^1 f(\alpha)(1 + \alpha)\, d\alpha = c \cdot \left(1 - \frac{\gamma^2}{1 + \gamma} - \frac{e^{\mu - \gamma}\mu}{1 + \gamma}\right) \leq c \cdot \left(1 - \frac{\gamma^2 + (1 + \mu - \gamma)\mu}{1 + \gamma}\right)$$
$$= c \cdot \left(1 - \frac{(\gamma - \mu)^2 + (1 + \gamma)\mu}{1 + \gamma}\right) \leq c \cdot (1 - \mu).$$

Theorem 1.8 *If α is chosen accordingly to the truncated exponential function f, as defined above, then the expected value of the resulting random α-schedule is bounded by $c \cdot Z_D$, so c times the optimum value.*

Proof From Eq. (15) and by Lemma 1.6 we have that

$$\begin{aligned} E_f[C_j^\alpha] &\leq t_j(0^+) + (c-1)\sum_{k\in N_1} \eta_k \cdot p_k + c\sum_{k\in N_2}(1-\mu_k)\cdot p_k + c\cdot p_j \\ &\leq c\cdot t_j(0^+) + c\sum_{k\in N_2}(1-\mu_k)p_k + c\cdot p_j = c\cdot\left(M_j^{LP} + \frac{p_j}{2}\right) \end{aligned}$$

where the last inequality follows from the definitions of η_k in the set N_1, so $\sum_{k\in N_1}\eta_k\cdot p_k \leq t_j(0^+)$, and the equality by Eq. (14).

This is a general theorem, so any function satisfying the hypothesis of Lemma 1.6 would define an upper bound of $c\cdot Z_D$.

The truncated exponential function (17) minimizes c in (19), since any function $\alpha \mapsto (c-1)\cdot e^\alpha$ verifies (18) with equality.

Theorem 1.9 *Let the density function be defined, for* $0 \leq a < b \leq 1$, *by*

$$f(\alpha) = \begin{cases} (c-1)\cdot e^\alpha & \text{if } \alpha \in [a,b] \\ 0 & \text{otherwise.} \end{cases} \tag{20}$$

Then the best bound satisfying Lemma 1.6 is reached for $a = 0$ *and* $b = \ln\frac{c}{c-1}$.

Proof Since it must define a density function over a probability distribution, it follows $(c-1)(e^b - e^a) = 1$, hence $c - 1 = \frac{1}{e^b - e^a}$. Considering Eq. (18) we have equality for $a = 0$, hence

$$\int_a^\eta f(\alpha)(1+\alpha-\eta)\,d\alpha \leq \int_0^\eta f(\alpha)(1+\alpha-\eta)\,d\alpha \leq (c-1)\cdot\eta \quad \text{for } \eta\in[0,1]$$

Thus we can focus on Eq. (19), and we want to minimize c.

$$\int_{\max\{\mu,a\}}^b f(\alpha)(1+\alpha)\,d\alpha \leq \int_\mu^b f(\alpha)(1+\alpha)\,d\alpha \qquad \text{for } a = 0$$

Then we have to find the smallest value of c in order to satisfy, for all $\mu\in[0,1]$ the inequality $\int_\mu^b f(\alpha)(1+\alpha)\,d\alpha \leq c\cdot(1-\mu)$. Hence

$$\frac{be^b - \mu e^\mu}{e^b - e^a} = \int_\mu^b \frac{f(\alpha)(1+\alpha)}{e^b - e^a}\,d\alpha \leq c\cdot(1-\mu) = \frac{e^b - e^a + 1}{e^b - e^a}(1-\mu), \tag{21}$$

so, in order to minimize c, we can derive over the variable μ and equate to 0, hence

$$\frac{(1+\mu)e^\mu}{e^b - e^a} = \frac{e^b - e^a + 1}{e^b - e^a} = c. \tag{22}$$

Then we plug in this value for c in (21), so

$$\frac{be^b - \mu e^\mu}{e^b - e^a} = \frac{(1+\mu)e^\mu}{e^b - e^a}(1-\mu) \implies \frac{be^b}{e^b - e^a} = \frac{(1+\mu-\mu^2)e^\mu}{e^b - e^a}.$$

Since we can multiply both sides for $e^b - e^a$ we have $be^b = (1+\mu-\mu^2)e^\mu$, which is not a-dependent. Since $c-1$ is minimized for $a=0$, plugging this value in (22) we have

$$\frac{(1+\mu)e^\mu}{e^b - 1} = \frac{e^b - 1 + 1}{e^b - 1} = c \implies b = \ln\left(\frac{c}{c-1}\right).$$

We show now why we have Eq. (16). We have $e^b = (1+\mu)e^\mu$ that equals $b = \mu + \ln(1+\mu)$. At the same time, if we substitute e^b and e^a in (21) it follows

$$b(1+\mu)e^\mu = (1+\mu-\mu^2)e^\mu \quad \text{that is} \quad b(1+\mu) = 1+\mu-\mu^2$$

$$\text{hence} \quad b = \frac{1+\mu-\mu^2}{1+\mu} = 1 - \frac{\mu^2}{1+\mu} \implies 1 - \frac{\mu^2}{1+\mu} = \mu + \ln(1+\mu)$$

and we have the equation for $c = \frac{e^b}{e^b - 1}$, by the result found before, namely $e^b = (1+\mu)e^\mu$, hence it follows

$$c = \frac{e^b}{e^b - 1} = \frac{(1+\mu)e^\mu}{(1+\mu)e^\mu - 1} = \frac{1+\mu}{1+\mu-e^{-\mu}}.$$

We conculde this section just pointing out that there is an easy way to derandomize the algorithm, see Sect. 4.3.

5.4 Bounds for Job-Dependent α_j Values

Following the previous section, we will now prove a better bound for the case of job-dependent α_j and we will present a different deterministic algorithm running in $O(n^2)$ time, see [15].

Let us have a probability distribution over (0, 1] with density function

$$g(\alpha) = \begin{cases} (c-1)\cdot e^\alpha & \text{if } \alpha \le \gamma + \ln(2-\gamma) = \delta \\ 0 & \text{otherwise} \end{cases} \tag{23}$$

where $\gamma \approx 0.4835$ is a solution to the equation $e^{-\gamma} + 2\gamma + \ln(2-\gamma) = 2$ and $c = 1 + \frac{1}{(2-\gamma)\cdot e^\gamma - 1} < 1.6853$. Let the α_j's be chosen pairwise independently from a probability distribution over (0, 1] with density function (23).

We start by proving a Lemma similar to Lemma 1.6, where the following inequalities (24) and (25) bound the delay of job j caused, respectively, by jobs in N_1 and in N_2.

Lemma 1.7 *The function g defined in* (23) *is a density function with the following properties:*

$$\int_0^{\eta} g(\alpha)\cdot(1+\alpha-\eta)\,d\alpha \le (c-1)\cdot\eta \quad \textit{for all } \eta\in[0,1], \tag{24}$$

$$(1+E_g[\alpha])\int_{\mu}^{1} g(\alpha)\,d\alpha \le c\cdot(1-\mu) \quad \textit{for all } \mu\in[0,1]. \tag{25}$$

where $E_g[\alpha]$ is the expected value of the random variable α distributed according to g.

Proof In order to have a density function we must have $\delta = \ln\left(\frac{c}{c-1}\right)$, since the calculation is identical to that in Lemma 1.6. We do not prove inequality (24) either because, also here, the proof is identical to the one in Lemma 1.6. In order to prove inequality (25) the idea is to compute first $E_g[\alpha]$, and then to show the correctness. So we have, by the definition of δ, that $E_g[\alpha]$ equals

$$\int_0^1 \alpha\, g(\alpha)\,d\alpha = (c-1)\int_0^{\delta}\alpha e^{\alpha}\,d\alpha = (c-1)\cdot[\delta e^{\delta}-e^{\delta}+1] = c\delta-1.$$

The inequality is certainly true for $\mu\in(\delta,1]$, while for $\mu\in[0,\delta]$ we have

$$\begin{aligned}
(1+E_g[\alpha])\int_{\mu}^{1} g(\alpha)\,d\alpha &= c\delta\cdot(c-1)\int_{\mu}^{\delta} e^{\alpha}d\alpha\\
&= ce^{-\gamma}\left((2-\gamma)e^{\gamma}-e^{\mu}\right)\\
&= c\left(2-\gamma-e^{\mu-\gamma}\right)\\
&\le c\left(2-\gamma-(1+\mu-\gamma)\right) = c\,(1-\mu).
\end{aligned}$$

We can now state and proof the following.

Theorem 1.10 *If α_j's are chosen pairwise independently from a probability distribution over* $(0,1]$ *with density function g as defined in* (23)*, then the expected value of the resulting α_j-schedule is bounded by c times the optimum LP value.*

Proof We start by considering a fixed choice of α_j in such a way to bound the conditional expectation of C_j^{α}. Then we integrate over all possible choices to have the desired final bound. For a given job j and a fixed α_j value, we have, according to Eq. (15) and by the property (24) stated in Lemma 1.7 that $E_g[C_j^{\alpha}|\alpha_j]$ is less or equal than

$$
\begin{aligned}
&t_j(0^+) + (1+\alpha_j)\cdot p_j + (c-1)\sum_{k\in N_1}\eta_k\cdot p_k + \sum_{k\in N_2:\alpha_j>\mu_k(j)}(1+E_g[\alpha_k])\cdot p_k\\
&\quad\le c\cdot t_j(0^+) + (1+E_g[\alpha_1])\sum_{k\in N_2:\alpha_j>\mu_k(j)}p_k + (1+\alpha_j)\cdot p_j,
\end{aligned}
$$

where the second inequality follows by (13) and by the equality of the expectations, that is, $E_g[\alpha_k] = E_g[\alpha_1]$ for all $k \in N$. Using now the Eq. (14) for the midpoints and property (25) of Lemma 1.7, we have

$$
\begin{aligned}
E_g[C_j^\alpha] &\le c\cdot t_j(0^+) + (1+E_g[\alpha_1])\sum_{k\in N_2}p_k\int_{\mu_k}^1 g(\alpha_j)\,d\alpha_j + (1+E_g[\alpha_j])\cdot p_j\\
&\le c\cdot t_j(0^+) + \sum_{k\in N_2}(1-\mu_k)\cdot p_k + c\cdot p_j = c\cdot\left(M_j^{LP}+\frac{p_j}{2}\right).
\end{aligned}
$$

The final result follows easily from the linearity of the expectations.

To describe a deterministic algorithm, we need to know how many possible (α_j)-schedules there are. We know that the number of possible orderings of n jobs is $n! = 2^{O(n\log n)}$, but the number of (α_j)-schedules is bounded.

Lemma 1.8 *The maximum number of (α_j)-schedules is at most 2^{n-1} and the bound is tight.*

Proof Let q_j denote the number of different pieces of the job j in the LP schedule, it means q_j is one more than the number of preemptions of job j. Since in the LP schedule there are at most $n-1$ preemptions, it follows $\sum_{j=1}^n q_j \le 2n-1$. The number of (α_j)-schedules, denoted by s, is bounded by the q_j's. So $s = \prod_{j=1}^n q_j = \prod_{j=2}^n q_j$ since the job with highest $\frac{w_j}{p_j}$ ratio is never preempted. Hence, just by applying the arithmetic–geometric mean inequality, we have

$$
s = \prod_{j=2}^n q_j \le \left(\frac{\sum_{j=2}^n q_j}{n-1}\right)^{n-1} \le 2^{n-1}.
$$

The proof of tightness can be easily seen in the instance with $p_j = 2$ and $r_j = n-j$ for all j, then we get $q_j = 2$ for $j = 2,\dots,n$.

The number of possible (α_j)-schedules is exponential, so enumerating all of them does not give a polytime algorithm. Instead, we use a different method, called of conditional probabilities.

Proposition 1.8 *There is a deterministic algorithm that is based on the 1.6853-approximation algorithm, with the same performance guarantee and running time of $O(n^2)$.*

Proof Let us denote the right-hand side of inequality (15) by $RHS(\boldsymbol{\alpha})$ and let $\boldsymbol{\alpha} = (\alpha_j)$. Then denoting $\sum w_j RHS_j(\boldsymbol{\alpha}) := V(\boldsymbol{\alpha})$ we already showed, for $c < 1.6853$, that $E_g[\sum_j w_j C_j^{\boldsymbol{\alpha}}] \le E_g[V(\boldsymbol{\alpha})] \le c \cdot Z_D$. Let us denote, for each job $j \in N$, by $\mathcal{Q}_j = \{\mathcal{Q}_{j1}, \dots, \mathcal{Q}_{jq_j}\}$ the set of intervals for α_j corresponding to the q_j pieces of job j in the LP schedule. Let us consider now the jobs, one by one, in arbitrary order, say, without loss of generality, $j = 1, \dots, n$. Assume that, at the step j of the derandomized algorithm, we identified the intervals $Q_1^d \in \mathcal{Q}_{j1}, \dots, Q_{j-1}^d \in \mathcal{Q}_{j-1}$ so that

$$E_g[V(\boldsymbol{\alpha})|\alpha_i \in Q_i^d \text{ for } i = 1, \dots, j-1] \le c \cdot Z_D.$$

By the use of the conditional expectations, we have that the left-hand side of this inequality is

$$\begin{aligned} &E_g[V(\boldsymbol{\alpha})|\alpha_i \in Q_i^d \text{ for } i = 1, \dots, j-1] \\ &\quad = \sum_{l=1}^{q_j} \Pr\left(\alpha_j \in Q_{jl}\right) \cdot E_g[V(\boldsymbol{\alpha})|\alpha_i \in Q_i^d \text{ for } i = 1, \dots, j-1 \text{ and } \alpha_j \in Q_{jl}]. \end{aligned}$$

Since $\sum_{l=1}^{q_j} \Pr\left(\alpha_j \in Q_{jl}\right) = 1$, there exists at least one interval $Q_{jl} \in \mathcal{Q}_j$ such that

$$\begin{aligned} &E_g[V(\boldsymbol{\alpha})|\alpha_i \in Q_i^d \text{ for } i = 1, \dots, j-1 \text{ and } \alpha_j \in Q_{jl}] \\ &\quad \le E_g[V(\boldsymbol{\alpha})|\alpha_i \in Q_i^d \text{ for } i = 1, \dots, j-1]. \end{aligned} \tag{26}$$

Therefore we just have to identify an interval $Q_j^d = Q_{jl}$ such that the inequality (26) is satisfied. Then we may conclude that

$$E_g\left[\sum_{h \in N} w_h C_j^{\boldsymbol{\alpha}} |\alpha_i \in Q_i^d, i = 1, \dots, j\right] \le E_g[V(\boldsymbol{\alpha})|\alpha_i \in Q_i^d, i = 1, \dots, j] \le c \cdot Z_D.$$

So we determined an interval Q_j^d for every job $j = 1, \dots, n$. Then we have that the (α_j)-schedule is the same for all $\boldsymbol{\alpha} \in Q_1^d \times \cdots \times Q_n^d$. So the deterministic objective value is now

$$\sum_{j \in N} w_j C_j^{\alpha} \le E_g[V(\boldsymbol{\alpha})|\alpha_i \in Q_i^d, i = 1, \dots, n] \le E_g[V(\boldsymbol{\alpha})] \le c \cdot Z_D.$$

For every $j = 1, \dots, n$ we need to evaluate $O(n)$ terms for checking if Q_j^d satisfies inequality (26), where each evaluation can be computed in constant time. By Lemma 1.8 there are at most $2n - 1$ intervals, hence the derandomized algorithm runs in $O(n^2)$ time.

5.5 Bad Instances for the LP Relaxations

In this section, we will present a family of bad instances, showing that the ratio between the optimum value of the $1|r_j|\sum w_j C_j$ problem and the lower bounds for the LP is arbitrarily close to $\frac{e}{e-1} > 1.5819$.

We can define these instances I_n, where n is the number of the jobs, such that there is one large job denoted by n, with processing time $p_n = n$, weight $w_n = \frac{1}{n}$, and release date $r_n = 0$. The other $n-1$ jobs, denoted $j = 1, \dots, n-1$, are small, they have 0 processing time, release date $r_j = j$, and weight $w_j = \frac{1}{n(n-1)}\left(1 + \frac{1}{n-1}\right)^{n-j}$. If zero processing times are not allowed, it is just sufficient to impose $1/k$ as processing time of the small jobs and multiply all the processing times and the release dates by k, and then let k tend to infinity. But for the sake of simplicity we assume that zero processing times are possible. In the LP solution job n starts at time 0 and it is preempted by each of the small jobs, hence it follows that $M_n^{LP} = \frac{n}{2}$ and $M_j^{LP} = r_j = j$ for the small jobs $j = 1, \dots, n-1$. Thus we can calculate the objective value, so $Z_R = \left(1 + \frac{1}{n-1}\right)^n - \left(1 + \frac{1}{n-1}\right)$.

Consider now an optimal nonpreemptive schedule C^* and let $k = \lfloor C_n^* \rfloor$, it means that k is the number of small jobs processed before n. It would be optimal to process these small jobs $j = 1, \dots, k$ at their release dates and start processing n at $r_k = k$. Then the remaining jobs would be processed at time $k + n$, hence after job n. Calling the obtained schedule C^k we have $C_j^k = j$ for $j \le k$ and $C_j^k = n + k$ otherwise.

The objective value of C^k is $\left(1 + \frac{1}{n-1}\right)^n - \frac{1}{n-1} - \frac{k}{n(n-1)}$.

The optimum schedule is C^{n-1} with value $\left(1 + \frac{1}{n-1}\right)^n - \frac{1}{n-1} - \frac{1}{n}$. Since n grows large we have that the optimum nonpreemptive cost tends to e, while the LP objective value is closer to $e - 1$, thus we have the desired ratio.

6 Approximations for $1|r_j|\sum C_j$

In this section we will study the problem $1|r_j|\sum C_j$ that can be seen as the problem of Sect. 5 with uniform weights. Chekuri et al. in [10] studied some α-schedule in a different way from the previous sections; we will present their result, namely an $\frac{e}{e-1}$-approximation algorithm for $1|r_j|\sum C_j$. We will also show some examples showing the tightness of the results, see [10] and the example of Torng and Uthaisombut in [44].

6.1 $\frac{e}{e-1}$-Approximation Algorithm

We already proved in Theorem 1.5 that we have an $f(\alpha)$-approximation algorithm, for $f(\alpha) = \max\{1 + \frac{1}{\alpha}, 1 + 2\alpha\}$. In the following, we prove a similar statement in a different way and then we refine the analysis by the use of some random α-points, in such a way to create the structure for the main result of this chapter.

Let C_i^P and C_i^α be the completion times of job i in a preemptive schedule P and in a nonpreemptive α-schedule S^α derived from P, respectively.

Theorem 1.11 *Let us consider the problem* $1|r_j|\sum C_j$. *For any* $\alpha \in (0, 1]$ *an* α*-schedule satisfies* $\sum C_j^\alpha \le \left(1 + \frac{1}{\alpha}\right) \sum C_j^P$.

Proof Given a preemptive schedule P, let us reindex the jobs in order of α-point, i.e., $t_1 \le t_2 \le \dots \le t_n$, and let $r_j^{max} = \max_{1 \le k \le j} r_k$. By maximality we have that $r_j^{max} \ge r_k$ for $k = 1, \dots, j$, hence

$$C_j^\alpha \le r_j^{max} + \sum_{k=1}^{j} p_k. \tag{27}$$

Since by r_j^{max} only an α-fraction of job j has finished we have $C_j^P \ge r_j^{max}$. We also know, by the ordering, that jobs $k \le j$ are processed for at least $\alpha \cdot p_k$ before C_j^P, hence $C_j^P \ge \alpha \sum_{k=1}^{j} p_k$. Substituting these two inequalities in (27) yields $C_j^\alpha \le \left(1 + \frac{1}{\alpha}\right) C_j^P$. Thus we just have to sum over all the jobs to obtain the statement.

In the following the conversion applies in general, so, in order to prove upper bounds on the performance ratio for the sum of completion times, we assume the preemptive schedule to be optimal. Hence we have a lower bound on any optimal nonpreemptive schedule.

We are now ready to define the main ideas on which the proof is based: let $S_i^P(\beta)$ denote the set of jobs that complete exactly a β-fraction of their processing time at time C_i^P, or the sum of the processing times of the jobs in this set (it will be clear case by case). Clearly $i \in S_i^P(1)$. We define τ_i as the idle time before C_i^P.

Lemma 1.9 *The completion time of job i in the preemptive schedule P equals* $C_i^P = \tau_i + \sum_{0 < \beta \le 1} \beta\, S_i^P(\beta)$.

Proof The completion time of job i corresponds to the sum of the idle time before C_i^P and the fractional processing times before C_i^P.

Lemma 1.10 *We can bound the completion time in the* α*-schedule by*

$$C_i^\alpha \le \tau_i + (1 + \alpha) \sum_{\beta \ge \alpha} S_i^P(\beta) + \sum_{\beta < \alpha} \beta\, S_i^P(\beta).$$

Proof We give a procedure converting the preemptive schedule P to a different schedule Q such that:

(c_1) jobs $1, \dots, i-1$ run nonpreemptively in that order,
(c_2) the remaining jobs can run preemptively
(c_3) C_i^Q satisfies the bound given in the lemma.

The proof of the lemma follows from the fact that $C_i^\alpha \le C_i^Q$. By Lemma 1.9 we have

$$C_i^P = \tau_i + \sum_{\beta<\alpha} \beta\, S_i^P(\beta) + \sum_{\beta\ge\alpha} \alpha\, S_i^P(\beta) + \sum_{\beta\ge\alpha} (\beta-\alpha)\, S_i^P(\beta). \tag{28}$$

Let us partition the set on jobs into $J^B := \bigcup_{\beta\ge\alpha} S_i^P(\beta)$ and $J^A = N - J^B$. So the Eq. (28) can be seen as the sum of:

(p_1) the idle time in S^P before C_i^P,
(p_2) the pieces of jobs in J^A running before C_i^P,
(p_3) the pieces of $j \in J^B$ running before t_j,
(p_4) the pieces of $j \in J^B$ running in $[t_j, C_i^P]$.

Let us denote by x_j the fraction of job j completed before C_i^P, that is the β for which $j \in S_i^P(\beta)$. So $(x_j - \alpha)\, p_j$ is just the fraction of j done in the interval $[t_j, C_i^P]$, hence we can write $\sum_{j\in J^B}(x_j-\alpha)\, p_j$ instead of $\sum_{\beta\ge\alpha}(\beta-\alpha)\, S_i^P(\beta)$. Let now $J^C = \{1,\dots,i\} =: [i]$ be a subset of J^B and consider P as an ordered list of pieces of jobs. Then for each $j \in J^C$ we:

- remove all pieces of the jobs running in $[t_j, C_i^P]$
- insert a piece of size $(x_j - \alpha)p_j$ at time t_j.

So for jobs $j = 1,\dots,i$ we have pieces of size $(x_j - \alpha)\, p_j$ in the correct order. Then we define a schedule from this list by scheduling the pieces of the jobs in the list order, respecting release dates. Obviously job i still completes at C_i^P since the pieces of jobs different by $(x_j - \alpha)\, p_j$ are moved later in time, so we do not add idle time. Then the algorithm extends the pieces of jobs by adding $p_j - (x_j - \alpha)\, p_j$, in such a way to have, for each job j, only one piece of size p_j. The pieces of j processed earlier, in total $\alpha\, p_j$, are replaced by idle time. The resulting schedule is the desired Q, in which the jobs $1,\dots,i$ are scheduled nonpreemptively for their processing time. Then the completion time of i is:

$$\begin{aligned} C_i^Q &= C_i^P + \sum_{j\in J^C} (p_j - (x_j-\alpha)\, p_j) \le C_i^P + \sum_{j\in J^B} (p_j - (x_j-\alpha)\, p_j) = \\ &= C_i^P + \sum_{\beta\ge\alpha} (1-\beta+\alpha)\, S_i^P(\beta) = \tau_i + (1+\alpha) \sum_{\beta\ge\alpha} S_i^P(\beta) + \sum_{\beta<\alpha} \beta\, S_i^P(\beta), \end{aligned}$$

where the last equation follows by substituting the value of C_i^P as in Lemma 1.9. The remaining pieces in the schedule are all from jobs in $N - J^C$, so we are done since it satisfies (c_1), (c_2), and (c_3).

We now restate the procedure used in Lemma 1.10 in such a way to define a conversion algorithm, called Q_α:

- let $J^B = \cup_{\beta\ge\alpha} S_i^P(\beta)$ and, for a fixed i, $J^C = \{1,\dots,i\} \subseteq J^B$,
- for each $j \in J^C$ remove all pieces of jobs running in $[t_j, C_i^P]$,
- insert a piece of size $(x_j - \alpha)\cdot p_j$ at the point corresponding to t_j,

- extend the piece of job j of size $(x_j - \alpha) \cdot p_j$ to a piece of size p_j,
- replace the other pieces of job j by idle time.

This algorithm has its full strength with a random choice of α.

Lemma 1.11 *Let α be drawn randomly from $(0, 1]$ over a probability distribution with density function f. Let $\delta = \max_{0<\beta\leq 1} \int_0^\beta \frac{1+\alpha-\beta}{\beta} f(\alpha)\, d\alpha$, then, for each job i we have $E[C_i^\alpha] \leq (1+\delta)\, C_i^P$.*

Proof By Lemma 1.10 we have

$$C_i^\alpha \leq \tau_i + (1+\alpha)\sum_{\beta\geq\alpha} S_i^P(\beta) + \sum_{\beta<\alpha} \beta S_i^P(\beta).$$

Hence choosing α accordingly to f, since τ does not depend on α we have $E[C_i(\alpha)] =$

$$\begin{aligned}
\int_0^1 f(\alpha)\, C_i^\alpha\, d\alpha &\leq \tau_i + \int_0^1 f(\alpha) \left((1+\alpha)\sum_{\beta\geq\alpha} S_i^P(\beta) + \sum_{\beta<\alpha} \beta\, S_i^P(\beta) \right) d\alpha \\
&= \tau_i + \sum_{0<\beta\leq 1} S_i^P(\beta) \left(\int_0^\beta (1+\alpha)\, f(\alpha)\, d\alpha + \int_\beta^1 \beta\, f(\alpha)\, d\alpha \right) \\
&= \tau_i + \sum_{0<\beta\leq 1} \beta\, S_i^P(\beta) \left(1 + \int_0^\beta \frac{1+\alpha-\beta}{\beta} f(\alpha)\, d\alpha \right) \\
&\leq \tau_i + \left(1 + \max_{0<\beta\leq 1} \int_0^\beta \frac{1+\alpha-\beta}{\beta} f(\alpha)\, d\alpha \right) \sum_{0<\beta\leq 1} \beta\, S_i^P(\beta) \\
&\leq \tau_i + (1+\delta) \sum_{0<\beta\leq 1} \beta\, S_i^P(\beta) \leq (1+\delta)\, C_i^P.
\end{aligned}$$

Then, by the linearity of expectations, we have the desired result.

With the next lemma we will prove some bound for specific density function over $(0, 1]$.

Theorem 1.12 *Let α be chosen accordingly to a specific probability distribution, then:*

1. *if α is chosen uniformly in $(0, 1]$ the expected approximation ratio is at most 2,*
2. *if α is drawn from $(0, 1]$ accordingly to the density function $f(\alpha) = \frac{e^\alpha}{e-1}$, then the expected approximation ratio is at most $\frac{e}{e-1} \approx 1.58$.*

Proof By Lemma 1.11 we want to bound the value $1+\delta$, so:

1. in this case we have $f(\alpha)=1$, from which it follows

$$\delta = \max_{0<\beta\le 1}\int_0^\beta \frac{1+\alpha-\beta}{\beta}\,d\alpha = \max_{0<\beta\le 1}\left(1-\beta+\frac{\beta}{2}\right) = \max_{0<\beta\le 1}\left(1-\frac{\beta}{2}\right),$$

   hence $\delta \le 1$, and it follows $1+\delta \le 2$.
2. in this case we have $f(\alpha)=\frac{e^\alpha}{e-1}$, hence

$$\begin{aligned}\delta &= \max_{0<\beta\le 1}\int_0^\beta \left(\frac{1-\beta+\alpha}{\beta}\right)\left(\frac{e^\alpha}{e-1}\right)d\alpha = \\ &= \max_{0<\beta\le 1}\frac{1}{\beta\,(e-1)}\left[e^\alpha-\beta\,e^\alpha+\alpha\,e^\alpha-e^\alpha\right]_0^\beta = \\ &= \max_{0<\beta\le 1}\frac{1}{\beta\,(e-1)}\left[e^\beta-1-\beta\,e^\beta+\beta+\beta\,e^\beta-e^\beta+1\right] = \\ &= \max_{0<\beta\le 1}\frac{1}{\beta\,(e-1)}\beta = \max_{0<\beta\le 1}\frac{1}{e-1}\end{aligned}$$

   so $1+\delta \le 1+\frac{1}{e-1}=\frac{e}{e-1}$.

As we proved in Lemma 1.5, for a given preemptive schedule, there are at most n combinatorially distinct values of α. Since each of them can be computed in $O(n)$ time and the fractional schedule can be computed in $O(n\log n)$ time we have the following.

Theorem 1.13 *There is a deterministic* $\frac{e}{e-1}$*-approximation algorithm for* $1|r_j|\sum C_j$ *running in* $O(n^2)$ *time.*

6.2 Tightness

Proposition 1.9 *There are instances in which the bound given by Theorem 1.11 is asymptotically tight.*

Proof Let $\epsilon>0$. In the desired instance there are $n+2$ jobs: job 1 such that $r_1=0$ and $p_1=1$, job 2 such that $r_2=\alpha-\epsilon$ and $p_2=\epsilon$, and job i, $i=3,\ldots,n+2$ such that $r_i=\alpha+\epsilon$ and $p_i=0$. The optimal preemptive completion time is $\alpha+n\,(\alpha+\epsilon)+1+\epsilon$, but the completion time of the derived nonpreemptive α-schedule is $\alpha+(\alpha+1)+n\,(1+\alpha)$. So for big n and small ϵ the ratio tends to $1+\frac{1}{\alpha}$.

Theorem 1.14 *There are instances in which the bound given by Theorem 1.12 is asymptotically tight.*

Proof In order to prove the statement, we just present the instances and give the results, without explaining in detail the results:

- Let us have $n+1$, jobs j, for $j \in [n]$, are such that $r_j = \delta < 1$ and $p_j = 0$, and job $n+1$ so that $p_{n+1} = 1$ and $r_{n+1} = 0$. The optimal nonpreemptive schedule has a completion time $C^* = n\,\delta + (1+\delta)$. There are only two combinatorially distinct values of α, so either $\alpha \leq \delta$ or $\alpha > \delta$. If α is chosen uniformly random, the algorithm output $C' = \delta\,(1+n) + (1-\delta)\,(1+\delta+n\,\delta)$. For $n \gg 1 \gg \delta$ the ratio between C' and C^* tends to 2.
- Let us have the following class of instances: $r_1 = 0, r_j = 1 + (n-j+1)\,\epsilon$ for $i = 2, \ldots, \lceil \frac{n}{e} \rceil - 1$ and $r_j = \sum_{k=j}^{n} \frac{1}{k} + (n-j)\,\epsilon$ for $j = \lceil \frac{n}{e} \rceil, \ldots, n$ and $p_1 = 1 + \epsilon$, $p_j = \epsilon$ for $j = 2, \ldots, n$. Then it follows that the algorithm computes $C' \geq n + \left(\frac{n\,(n+1)}{2}\right)\epsilon$, while the optimum value is

$$C^* < \left(\frac{e-1}{e}\right) n + \left(2 - \frac{1}{e}\right) + \left(\frac{n\,(n+1)}{2}\right)\epsilon + 1 + \left\lfloor \frac{n}{e} \right\rfloor \epsilon.$$

  For any integer $N > 5$ there is an integer $n \geq 6$ such that we have $N - 4 \leq n \leq N$, $0 < \epsilon < \frac{1}{n}$, and $\sum_{k=\lceil n/e \rceil}^{n} \frac{1}{k} < 1$. Hence for any $\delta > 0$ there exists some $\epsilon > 0$ that makes the ratio $\frac{C'}{C^*} \geq \frac{e}{e-1} - \delta$, so it can be made arbitrarily close to $\frac{e}{e-1}$.

7 Approximation for $1|r_j, prec|\sum w_j C_j$

We now consider the more general problem $1|r_j, prec|\sum w_j C_j$, so there is a partial order defined by "$\prec$", meaning that, whenever $j \prec k$, we need to complete job j before to start job k. It is obvious that for $j \prec k$ we can assume $r_j \leq r_k$, or more specifically $r_j + p_j \leq r_k$.

In this section we will first explain a general algorithm that is called list scheduling. Then we will present the recent result of Skutella, in [41], where he obtained a $\frac{\sqrt{e}}{\sqrt{e}-1}$-approximation algorithm. So we will give the LP relaxation used and we will show a "nice" trick used by the author.

Since the problems $1|r_j|\sum C_j$, so with $w \equiv 1$ and no precedence constraints, and $1|prec|\sum C_j$, so with $w = 1$ and $r = 0$, are strongly NP-hard also the problem $1|r_j, prec|\sum w_j C_j$ is strongly NP-hard. If we would allow preemptions we could solve optimally in polynomial time $1|r_j, pmtn|\sum C_j$, but even the problems $1|r_j, pmtn|\sum w_j C_j$ and $1|prec, pmtn|\sum C_j$, that is equivalent to $1|prec|\sum C_j$, are strongly NP-hard. We have to note that Bansal and Khot proved in [5] a bound on the possible approximability, that is, by using a stronger version of the Unique Games Conjecture, there are no $(2-\epsilon)$-approximation algorithm for the problem $1|prec|\sum w_j C_j$. Another interesting fact is the relation between approximability of $1|prec|\sum w_j C_j$ and the vertex cover problem, see [1, 2, 12].

7.1 LP Relaxation

We define a *list schedule* to be a schedule that processes the jobs as early as possible, with respect to the release dates, in a given order. This order is given by a total ordering that extends (if any) the partial order given by $\prec$. If we are in the preemptive case we can define the preemptive list scheduling such that we process at any point in the time the first available job, in this way we obtain a preemptive list schedule.

From the problem $1|r_j, prec|\sum w_j C_j$ we can obtain the preemptive relaxation $1|r_j, prec, pmtn|\sum w_j C_j$ for which it is known a 2-approximation algorithm, see [19].

Let $S \subseteq N$ be a set of jobs and recall, as we defined in the previous sections, that $p(S) = \sum_{j \in S} p_j$ and $r_{min}(S) = \min_{j \in S} r_j$. Then using the variables C_j for $j \in N$ we can define the relaxation:

$$\min \quad \sum_{j \in N} w_j C_j$$

subject to

$$C_j + p_k \leq C_k \qquad \text{for all } j \prec k \qquad \text{(s1)}$$

$$\frac{1}{p(S)} \sum_{j \in S} p_j C_j \geq r_{min}(S) + \frac{p(S)}{2} \qquad \text{for all } S \subseteq N \qquad \text{(s2)}$$

Any optimal solution C^* of the LP relaxation can be found in polynomial time, because, also if there are exponentially many constraints, they can be separated in polynomial time by efficient submodular minimization algorithm, see [14]. Then we have that $\sum w_j C_j^*$ is a lower bound for the sum of weighted completion times over the preemptive schedules. Let us now reindex the jobs in order to have

$$C_1^* \leq C_2^* \leq \cdots \leq C_n^* \qquad (29)$$

By the LP formulation it is possible that there are some jobs j and k such that $j \prec k$ but $C_j^* = C_k^*$, but, in order to avoid this degenerate case, we assume the processing time to be nonzero, i.e., $p_j > 0$ for $j \in N$.

Lemma 1.12 *Given the ordering of jobs obtained by the reindexing* (29) *and a double speed machine, if we apply the preemptive list scheduling algorithm we obtain a feasible preemptive schedule with completion times C' such that $C'_j \leq C_j^*$ for all $j \in N$.*

Proof To prove the theorem we define, for any fixed job $j \in N$, a specific subset of jobs $S = \{k : k \leq j\} \subseteq N$ satisfying:

1. $C'_k \leq C'_j$,
2. there is no idle time in $[C'_k, C'_j]$ by the preemptive list scheduling,
3. in $[C'_k, C'_j]$ only jobs $l \leq j$ are processed.

The idea is to prove a nice property of set S, namely that jobs in S are processed continuously in $I := [r_{min}(S), C'_j]$, so let $h \in S$ be the job with $r_h = r_{min}(S)$. By condition (s1) and since $p_j > 0$ we can assume that $l \prec h$ implies $r_l < r_h$. So the algorithm processes only jobs $l \preceq h \preceq j$ and it does not leave idle time in the interval $[r_h, C'_h]$. By the third condition on the set S, in $[C'_h, C'_j]$ only jobs $l \preceq j$ are processed and, by the second one, there is no idle time. Hence we have that in I there is no idle time and only jobs $l \preceq j$ are processed. Any job l processed in I, since $l \prec j$, is also completed in I, i.e., $C'_l \in I$. Hence C'_l satisfies all the conditions and, therefore, it is contained in S.

Hence, since the jobs run on a double speed machine, it follows that $C'_j = r_{min}(S) + \frac{p(S)}{2}$, from which we have

$$C^*_j \geq \frac{1}{p(S)} \sum_{k \in S} p_k C^*_k \geq r_{min}(S) + \frac{p(S)}{2} = C'_j,$$

since $k \preceq j$ implies $C^*_k \leq C^*_j$ and by the LP constraints (s2).

Theorem 1.15 *Given the LP lower bound $\sum w_j C^*_j$ of a preemptive schedule on the optimal total weighted completion time. Then, using a double speed machine we can obtain, in polynomial time, a preemptive list schedule whose total weighted completion time satisfies $\sum w_j C'_j \leq \sum w_j C^*_j$.*

Proof By the LP relaxation we can find in polynomial time an optimal solution C^*, we then apply the preemptive list scheduling algorithm on the order given by the relation (29) on a double speed machine. By Lemma 1.12 it follows $\sum w_j C'_j \leq \sum w_j C^*_j$.

7.2 2.54-Approximation Algorithm

In this section we will do list scheduling in order of α-points. The definition of α-point for schedules on regular speed machine is known, and it is easy to generalize the same parameter for a double speed machine: the α-point $t'_j(\alpha)$, or just t'_j for common α, of job j with respect to schedule S' is the first time when $\alpha \cdot \frac{p_j}{2}$ units of time of job j have been processed on the double speed machine. Let now S^α be the schedule obtained by processing jobs in order of increasing $t'_j(\alpha)$ on a regular speed machine. Since S' is feasible this order respects the precedence constraints. We denote, as usual, by C^α_j the completion time of job j in the list schedule S^α. To be precise we redefine also the parameter η for the double speed machine, so, for a fixed job k, $\eta_j(k)$ (or just η_j when job k is clear from the context) is the fraction of job j done on the double speed machine by time C'_k. We have

$$C'_k \geq \sum_{j \in N} \eta_j \frac{p_j}{2} \tag{30}$$

Lemma 1.13

$$C_k^\alpha \leq C_k' + \sum_{j:\eta_j \geq \alpha} \left(1 + \frac{\alpha - \eta_j}{2}\right) \cdot p_j.$$

Proof Let X be the subset of all jobs j scheduled by S_α not later than k for which in $[C_j^\alpha, C_k^\alpha]$ there is no idle time, clearly $k \in X$. Let us denote by l the first job scheduled in X, hence $s_l = r_l$, where s_l is the starting time of job l. We have, by the definition of the set X, that:

$$C_k^\alpha = r_l + \sum_{j \in X} p_j. \tag{31}$$

Considering now the schedule S', by feasibility, for every job $j \in X$ we have that $r_l \leq t_l' \leq t_j'$ and, by definition of η_j, that the double speed machine processes j in $[t_j', C_k']$ for exactly $\frac{1}{2}(\eta_j - \alpha)\, p_j$. Since a machine cannot schedule more than one job at any time we have

$$C_k' \geq r_l + \sum_{j \in X} \frac{\eta_j - \alpha}{2}\, p_j. \tag{32}$$

Now we just have to combine Eq. (31) with (32) in order to get

$$C_k^\alpha \leq C_k' + \sum_{j \in X} \left(1 + \frac{\alpha - \eta_j}{2}\right) p_j. \tag{33}$$

Since $t_j' \leq t_k'$ we have $\eta_j \geq \alpha$ for $j \in X$, then $X \subseteq \{j : \eta_j \geq \alpha\}$ and, by the nonnegativity of α, $\left(1 + \frac{\alpha - \eta_j}{2}\right)$ is at least $\frac{1}{2}$, hence we have

$$C_k^\alpha \leq C_k' + \sum_{j \in X} \left(1 + \frac{\alpha - \eta_j}{2}\right) p_j \leq C_k' + \sum_{j:\eta_j \geq \alpha} \left(1 + \frac{\alpha - \eta_j}{2}\right) \cdot p_j.$$

Theorem 1.16 *Let S' be a feasible preemptive schedule on a double speed machine and denote the completion time of job j in S', for every $j \in N$, by C_j'. It is possible to obtain, in polynomial time, a feasible nonpreemptive schedule on a regular speed machine with completion times C_j so that*

$$\sum_{j \in N} w_j C_j \leq \frac{\sqrt{e}}{\sqrt{e} - 1} \sum_{j \in N} w_j C_j'.$$

Proof In order to prove the theorem we define the value of α by a suitable probability distribution and then we schedule the jobs on the regular speed machine in order of α-points with respect to the schedule S'. Let $\alpha \in (0, 1]$ be drawn randomly according to the density function

$$f(\alpha) := \frac{e^{\alpha/2}}{2(\sqrt{e}-1)}.$$

For $0 \leq \eta \leq 1$, we have

$$\int_0^{\eta} f(\alpha)\left(1+\frac{\alpha-\eta}{2}\right)d\alpha = \int_0^{\eta} \frac{2-\eta+\alpha}{2}\, e^{\alpha/2}\, d\alpha = \frac{\eta}{2(\sqrt{e}-1)}.$$

By Lemma 1.13 and by inequality (30) it follows

$$\begin{aligned} E[C_k^{\alpha}] &\leq C'_k + \sum_{j\in N} p_j \int_0^{\eta_j} f(\alpha)\left(1+\frac{\alpha-\eta_j}{2}\right)d\alpha \\ &= C'_k + \frac{1}{\sqrt{e}-1}\sum_{j\in N} \eta_j \frac{p_j}{2} \leq \frac{\sqrt{e}}{\sqrt{e}-1} C'_k. \end{aligned}$$

Then, by linearity of expectations, it follows that, in the nonpreemptive list schedule in order of random α-points, the sum of weighted completion times is within a factor of $\frac{\sqrt{e}}{\sqrt{e}-1}$ times the sum of weighted completion times in S'. As we proved in Theorem 1.3 there are at most n combinatorially different values of α, each of them can be computed in $O(n)$ time, so in total the running time is $O(n^2)$.

Corollary 1.3 *For the scheduling problem $1|r_j, prec|\sum w_j C_j$, any optimal solution of the LP relaxation defined in Sect. 7.1 is at most a factor $\frac{\sqrt{e}-1}{\sqrt{e}}$ smaller than the total weighted completion time of an optimal schedule.*

Proof Let C' denotes the vector of completion times in the preemptive double speed machine schedule S' and let C^* be an optimal LP solution. Then by Theorems 1.15 and 1.16 the algorithm constructs a feasible schedule in which the sum of weighted completion times is bounded by

$$\sum_{j\in N} w_j C_j \leq \frac{\sqrt{e}}{\sqrt{e}-1}\sum_{j\in N} w_j C'_j \leq \frac{\sqrt{e}}{\sqrt{e}-1}\sum_{j\in N} w_j C^*_j.$$

We complete the proof just by observing that $\sum w_j C_j$ is an upper bound of the value for an optimal schedule.

8 Approximations for $P|r_j|\sum C_j$

In this section we will first show an easy 3-approximation algorithm for the problem $P|r_j|\sum C_j$, given by Chekuri et al. in [10]. In the same paper the authors developed the DELAY-LIST algorithm that produces, given any ρ-approximation for the weighted sum of completion times on a single machine, a $((1+\beta)\rho + (1+1/\beta))$-approximation guarantee for the general m-machines case. Note that the DELAY-LIST algorithm works even with precedence constraints. At the end, we will present the 2.83-approximation algorithm obtained by combining the DELAY-LIST and a list scheduling algorithm, see [10].

8.1 3-Approximation Algorithm

Let M be an instance for a nonpreemptive scheduling on m machines, then we can define a preemptive one-machine instance I on the same set of jobs with $p'_j = \frac{p_j}{m}$ and $r'_j = r_j$.

Lemma 1.14 *Let M^* be an optimal value for the instance M and I^* be an optimal value for I. Then $I^* \leq M^*$.*

Proof Let S^m be a feasible schedule for the input M. Then it is possible to convert S^m to a feasible schedule S^I for input I, without increasing the average completion time. Consider a time unit t sufficiently small and denote, for S^m, the $k \leq m$ jobs running at time t by $1, \ldots, k$. For each job $1, \ldots, k$ at time t, in S^I, we can run $\frac{1}{m}$ units of each jobs. It follows that, for any job j the completion time in S^I, C^I_j, is not greater than the completion time in S^m, denoted C^m_j.

Let S^I be an optimal preemptive schedule for I; then we can create a list schedule S^m by ordering the jobs by increasing value of C^I_j and scheduling them in that order, in a nonpreemptive way, respecting the release dates. Let C^*_j be the completion time of j in an optimal schedule for M, then, despite I could be a bad relaxation, we can derive a good approximation guarantee.

Lemma 1.15 *Given a list of jobs in order of increasing C^I_j, let S^m be the nonpreemptive list schedule generated, then we have $\sum C^m_j \leq \left(3 - \frac{1}{m}\right) \sum C^*_j$.*

Proof We can assume (by reindexing) that the jobs are ordered by completion times in S^I, so $C^I_1 \leq C^I_2 \leq \cdots \leq C^I_n$, i.e. job j is the j-th job completed in S^I. Then we have two easy bounds: $C^I_j \geq r'_j + p'_j$ and

$$C^I_j \geq \sum_{k=1}^{j} p'_k = \sum_{k=1}^{j} \frac{p_k}{m}. \tag{34}$$

We can derive another trivial bound after defining $r_j^{max} := \max_{k\in[j]} r_k$, namely $C_j^I \geq r_j^{max}$. By definition we know that all the jobs $k \leq j$ have been released at time r_j^{max}, so we can bound the completion time of job j in S^m:

$$C_j^m \leq r_j^{max} + \sum_{k=1}^{j-1} \frac{p_k}{m} + p_j \leq C_j^I + C_j^I + p_j\left(1 - \frac{1}{m}\right), \tag{35}$$

by inequality (34) and since $C_j^I \geq r_j^{max}$. Then if we sum (35) over all jobs, by Lemma 1.14, we get

$$\sum_j C_j^m \leq 2 \cdot \sum_j C_j^I + \left(1 - \frac{1}{m}\right)\sum_j p_j \leq \left(3 - \frac{1}{m}\right)\sum_j C_j^* \tag{36}$$

since we know $\sum C_j^* \geq \sum p_j$.

8.2 The DELAY-LIST Algorithm

In this section we first sketch the DELAY-LIST algorithm and then we give a detailed explanation.

We can derive a list from the one-machine schedule by ordering the jobs in order of nondecreasing completion times. The precedence constraints prohibit the complete parallelization, so, in order to benefit from the parallelism, we may process jobs out-of-order. A job is scheduled in *order* whenever it is scheduled as first of the list, otherwise it is scheduled *out-of-order*. If the p_j's are identical we could schedule jobs out-of-order without any problem, otherwise a job could delay "better" jobs that would become available soon, so we have to balance this effect. We will schedule jobs out-of-order only if there is enough idle time to justify this change in the list.

We start by giving some definitions that will be used throughout this section. For any job j we can define recursively the value of the *smallest feasible completion time* of j, namely κ_j, as $r_j + p_j$ if the job has no predecessors and $p_j + \max\{r_j, \max_{i \prec j} \kappa_i\}$ otherwise. A path P from i to j is called *critical path to j* if $p(P) = \kappa_j$. If we fix a schedule S we can denote by q_j^S the time at which job j is ready (released and all predecessors completed) and by s_j^S the starting time of j in the schedule S.

We can now describe the algorithm where, for the sake of simplicity, we assume p_j to be integers and to have discrete time. In the algorithm the idle time is *charged* on the jobs, in such a way to have a rule to decide when to schedule a job out-of-order. Let $\beta > 0$ be a fixed constant and suppose the jobs are ordered on a list, then at each time step t the algorithm applies one of the following:

(a) if M is an idle machine and the first job on the list j is ready at time t, then schedule j on M and charge all uncharged idle time in (q_j, s_j) to j,

(b) if M is an idle machine and the first job on the list j is not ready at time t, then we consider the first ready job k in the list: if there is at least $\beta\, p_k$ uncharged idle time among all machine, then we schedule k and we charge $\beta\, p_k$ idle time on k,

(c) if either there is no idle time or the previous cases do not apply we merely increment t.

For any job j we partition $N \setminus \{j\}$ in $A_j = \{k \text{ after } j \text{ in the list}\}$ and in $B_j = \{k \text{ before } j \text{ in the list}\}$. We also denote by O_j the set of jobs that are after j on the list but are scheduled before job j, and we know $O_j \subseteq A_j$.

We can see that the algorithm considers only the first ready job in the list, even if jobs ordered after it in the list are ready.

Given a schedule S, for each job j, we can define a path $P'_j = k_1, \ldots, k_l = j$, where k_i is the largest completion time predecessor of k_{i+1} and ties are broken arbitrarily. Obviously job k_1 does not have predecessors which satisfy the condition.

Lemma 1.16 *Given a schedule S, we can find the path P'_j. The jobs in P'_j define a disjoint set of time intervals $(0, r_{k_1}]$, $(s_{k_1}, C_{k_1}]$, $\ldots$, $(s_{k_l}, C_{k_l}]$. Let κ'_j be the sum of lengths of these intervals, then $\kappa'_j \leq \kappa_j$.*

Proof It is easy to see that $s_k = \max\{r_k, C_{k-1}\}$ cannot be greater than $\max\{r_k, \max_{i \prec k} \kappa_i\}$, then, by definition of κ_i, we have the desired result.

Lemma 1.17 *The algorithm charges on each job j at most $\beta\, p_j$ time.*

Proof By the algorithm if case (b) applies we are done. So suppose j is ready at q_j and was not scheduled before according to case (b), then the idle time in (q_j, s_j) is not greater than $\beta\, p_j$. For this proof we assume the time to be continuous, so j is scheduled at the first time when $\beta\, p_j$ units of idle time have been accumulated, otherwise the algorithm might charge some extra idle time by the integrality.

Lemma 1.18 *For each job j, there is no uncharged idle time in the interval (q_j, s_j). Moreover all the idle time is charged to jobs in B_j.*

Proof By the algorithm no jobs in A_j are scheduled in (q_j, s_j) since j was ready at q_j, hence the idle time is charged on $B_j \cup \{j\}$. Since j is ready at q_j, but started only at s_j, by cases (a) and (b) of the algorithm there is no uncharged idle time to j.

Lemma 1.19 *Let j be any job, the idle time in $(0, s_j)$ charged to jobs in A_j is at most $m\,(\kappa'_j - p_j)$. Hence $p(O_j) \leq m\, \frac{(\kappa'_j - p_j)}{\beta} \leq m\, \frac{\kappa_j - p_j}{\beta}$.*

Proof Let us consider job $k \in B_j$ and let job k' be ready at the completion time of job k, that is $q_{k'} = C_k$. Since k is before j in the list, then it follows $A_j \subset A_k$, then by Lemma 1.18 in the interval (C_k, s'_k) no idle time is charged on A_j. Hence the idle time of jobs in A_j is accumulated in the intersection between the P'_j intervals and $(0, s_j)$, that is bounded by $m\,(\kappa'_j - p_j)$, since there are m machines and

$|(\cup P'_j) \cap (0, s_j)| \leq \kappa'_j - p_j$. Since $O_i \subseteq A_i$ the total idle time charged to O_i is at most $\frac{1}{\beta}$ times the possible idle time charged on A_i, then by Lemma 1.16 we have $p(O_i) \leq \frac{m\,(\kappa'_j - p_j)}{\beta} \leq \frac{m\,(\kappa_j - p_j)}{\beta}$.

Theorem 1.17 *Given a list the algorithm DELAY-LIST produces a schedule S, so that, for each job j, we have*

$$C_j \leq \frac{(1+\beta)p(B_j)}{m} + \left(1 + \frac{1}{\beta}\right)\kappa'_j - \frac{p_j}{\beta}.$$

Proof Let us consider job j, then we can partition the interval $T = (0, C_j)$ in $T_1 = \{\text{intervals defined by } P'_j\}$ and $T_2 = T - T_1$, and we define t_1 and t_2 as the sum, respectively, of the intervals in T_1 and T_2. By definition we have $t_1 = \kappa'_j \leq \kappa_j$. By Lemma 1.18 the idle time in T_2 is charged on jobs in the set B_j and the running jobs are from $B_j \cup O_j$, then by Lemma 1.17 we have

$$C_j = t_1 + t_2 \leq \kappa'_j + \frac{\beta\, p(B_j) + p(B_j) + p(O_j)}{m} \leq$$
$$\leq \kappa'_j + \frac{(\beta+1)\, p(B_j)}{m} + \frac{\kappa'_j - p_j}{\beta} = \kappa'_j \left(1 + \frac{1}{\beta}\right) + \frac{(\beta+1)\, p(B_j)}{m} - \frac{p_j}{\beta},$$

where the second inequality follows from Lemma 1.19.

Corollary 1.4 *Let S^I be a single machine schedule. Then, running the DELAY-LIST algorithm with the list generated by S^I, it outputs the schedule S^m such that*

$$C_j^m \leq \frac{C_j^I}{\beta}(1+\beta) + \left(1 + \frac{1}{\beta}\kappa_j\right).$$

Proof In S^1 all jobs in B_j are scheduled before j, hence $p(B_j) \leq C_j^I$. If we plug this inequality in Theorem 1.17 and by Lemma 1.16 we have

$$C_j^m \leq (1+\beta)\frac{C_j^I}{m} + \left(1 + \frac{1}{\beta}\right)\kappa_j.$$

The DELAY-LIST algorithm needs a one-machine schedule. So we see some relations between schedules with different machines. Let C_{OPT}^m be the total weighted completion time of an optimal m-machine schedule.

Lemma 1.20 *Given a single machine and a m-machine formulation we have $m \cdot C_{OPT}^m \geq C_{OPT}^I$, where C_{OPT}^I is the total weighted completion time for an optimal single machine schedule.*

Proof Given a schedule S^m on m machines with total weighted completion time C^m we can construct a single machine schedule S^I with $C^I \leq m\, C^m$. Let us order the jobs by S^m completion time and use this order for S^I. By the release dates there could be some idle time. Let us denote the sum of completion times of the jobs finishing before j, included j, by P and the total idle time in S^m before C_j^m by ID. Then $m\, C_j^m \geq P + ID$. Since P is the sum of all jobs before j in S^I and the idle time in S^I can be charged to idle time in S^m we get $C_j^I \leq P + ID \leq m\, C_j^m$.

Lemma 1.21 $C_{OPT}^m \geq \sum w_j C_j = C_{OPT}^\infty$.

Proof By definition $C_j^m \geq \kappa_j$, hence summing over j we get the inequality. If we have an unbounded number of machines, any job j can be scheduled at the earliest available time, so it will finish at κ_j, hence we have the equality.

Theorem 1.18 *Suppose to have a ρ-approximation algorithm for minimizing the sum of completion times in a one-machine problem, then DELAY-LIST gives an m-machine schedule within a factor of $(1+\beta)\,\rho + \left(1 + \frac{1}{\beta}\right)$ of an optimal m-schedule.*

Proof Let S^I be a one-machine schedule within a factor of ρ of the optimal one-machine schedule, then $\sum w_j C_j^I = C^I \leq \rho\, C_{OPT}^I$. By Corollary 1.4 the algorithm DELAY-LIST creates a schedule such that

$$
\begin{aligned}
C^m &= \sum w_j C_j^m \leq \sum w_j \left((1+\beta)\, \frac{C_j^I}{m} + \left(1 + \frac{1}{\beta}\right) \kappa_j \right) \\
&= \frac{1+\beta}{m} \sum w_j C_j^I + \left(1 + \frac{1}{\beta}\right) \sum w_j \kappa_j .
\end{aligned}
$$

Then by Lemmas 1.20 and 1.21 we get

$$
\begin{aligned}
C^m &\leq \frac{(1+\beta)\,\rho\, C_{OPT}^I}{m} + \left(1 + \frac{1}{\beta}\right) C_{OPT}^\infty \\
&\leq \left((1+\beta)\,\rho + \left(1 + \frac{1}{\beta}\right) \right) C_{OPT}^m .
\end{aligned}
$$

The power of the algorithm derives from the fact that the bounds are given job-by-job, hence we can apply the algorithm even for more general classes of metric.

8.3 2.83-Approximation Algorithm

We will derive a 2.83-approximation algorithm for the sum of completion times on m machines with release dates but no precedence constraints. We will follow the proof of Chekuri et al. see [10], that works by combining a list scheduling and the DELAY-LIST algorithms.

Let M and I be defined as in Lemma 1.14, and let C_j^I be the completion time of job j in an optimal schedule I^*.

Lemma 1.22 *Let S^m be the schedule obtained by applying the DELAY-LIST algorithm on I^* with parameter β. Denoting the completion time of job j in S^m by C_j^m, then it follows*

$$\sum C_j^m \leq (2+\beta)\, C_j^* + \frac{1}{\beta} \sum r_j.$$

Proof Let us consider job j. Trivially, since we do not have any precedence constraint, $\kappa_j = r_j + p_j$. By Lemma 1.16 and Theorem 1.17 we have that $C_j^m \leq (1+\beta)\, \frac{p(B_j)}{m} + \left(1 + \frac{1}{\beta}\right) \kappa_j - \frac{p_j}{\beta}$. Since the jobs are ordered by the single machine completion time we have $\frac{p(B_j)}{m} \leq C_j^I$. So combining these inequalities we get $C_j^m \leq (1+\beta)\, C_j^I + p_j + \frac{r_j}{\beta}$. Summing over all j we have

$$\sum C_j^m \leq (1+\beta) \sum C_j^I + \sum p_j + \frac{1}{\beta} \sum r_j \leq (2+\beta) \sum C_j^* + \frac{1}{\beta} \sum r_j,$$

because $\sum C_j^I$ and $\sum p_j$ are lower bounds on the optimal value.

Theorem 1.19 *For any m-machine instance, applying either list scheduling or DELAY-LIST for an appropriate β we obtain a schedule S^m such that $\sum C_j^m \leq 2.83 \sum C_j^*$. Moreover, it runs $O(n \log n)$ time.*

Proof By Eq. (36) we know $\sum C_j^m \leq 2 \sum C_j^I + \sum p_j$.

- if α is such that $\sum p_j \leq \alpha \sum C_j^*$, then the list scheduling algorithm has an approximation guarantee of $(2+\alpha)$,
- if $\sum p_j > \alpha \sum C_j^*$, since we know that $\sum C_j^* \geq \sum r_j + p_j$, then we obtain $\sum r_j \leq (1-\alpha) \sum C_j^*$. So, by Lemma 1.22, we get

$$\sum C_j^m \leq \left(2 + \beta + \frac{1-\alpha}{\beta}\right) \sum C_j^*.$$

Given α we can choose β so that we minimize $\min\{2+\alpha, 2+\beta+\frac{1-\alpha}{\beta}\}$. Choosing $\beta = \sqrt{1-\alpha}$, we have that $\min\{2+\alpha, 2+2\sqrt{1-\alpha}\}$ is minimized by $\alpha = 2\sqrt{2}-2$ and it gives an approximation guarantee of $2\sqrt{2} \approx 2.83$. So the algorithm chooses $\beta = \sqrt{3-2\sqrt{2}}$ and runs both the algorithms, keeping, at the end, the best schedule. The running time of both algorithms is $O(n \log n)$, so the final schedule can be computed in $O(n \log n)$ time.

9 Approximation for $P|d_{ij}|\sum w_jC_j$

In this chapter we will explain the list scheduling algorithm for parallel machines, defining the job-driven algorithm that will be used. Then we will show the proof for the 4-approximation algorithm for the problem $P|d_{ij}|\sum w_jC_j$ given by Queyranne and Schulz in [35], for which we will present the LP relaxation used. At the end we will give an example that proves the tightness of this algorithm, see [35].

9.1 List Scheduling Algorithms

One of the easiest methods to find approximate solutions for parallel machines scheduling problems consists of list scheduling algorithms. Given a list satisfying the precedence constraints, these algorithms schedule, when one of the m machines becomes idle, the first available job. This approach can cause some troubles because we may delay the schedule of jobs with large weight that may be available soon, by scheduling less important jobs. Hence sometimes adding some idle time would be useful to obtain a better approximation. We provide here an easy example to show this problem:

Example 1.1 Let us have a single machine instance, with jobs j and k for the problem $1|r_j|\sum w_jC_j$. For $q \geq 2$ assume: $p_j = q, r_j = 0$ and $w_j = 1$ for job j and $p_k = 1, r_k = 1$ and $w_k = q^2$ for job k. The optimal schedule leaves the machine idle in $[0, 1)$, then processes k and after j, so to have optimum value of $2q^2 + q + 2$. The list scheduling algorithm would process before j and after k, in such a way the objective value is $q^3 + q^2 + q$. Hence, for q big, we have unbounded performance ratio.

This example could be generalized also with precedence constraints. So the idea is to find a different way of list scheduling the jobs.

A simple strategy is to consider the jobs one by one in the given list order without altering this order, so we have a specific job-driven list scheduling. Given a list, that extends the partial order defined by precedence constraints, the algorithm works as follows:

1. assume the list to be $L = l_1, \ldots, l_n$, the machines to be empty and define the machine completion times $\Gamma_h := 0$ for $h = 1, \ldots, m$.
2. for j from l_1 to l_n define the starting and the completion times of j and assign j to a machine:
   a. $s_j := \max\left(\max\{C_i + d_{ij} : (i, j) \in A\}, \min\{\Gamma_h : h = 1, \ldots, m\}\right)$ is the starting time of job j and $C_j := s_j + p_j$ is its completion time,
   b. take a machine h with $\Gamma_h \leq s_j$ and assign job j to it. Then update the completion time of machine h, hence $\Gamma_h := C_j$.

We could use many rules for the choice of the machine to which assign a job, but for the purposes of the analysis we use a general one. Also the definition of the list L could be done in many ways, but we consider the list generated by sorting the jobs in order of nondecreasing completion times on a relaxation of the problem.

9.2 4-Approximation Algorithm

The precedence constraints define a partial order, hence a directed acyclic graph $D = (N, A)$ where N is the set of jobs and A represents the precedence constraints, i.e., we have $(i, j) \in A$ if $i \prec j$. In order to study the problem we need to define what is the parameter d_{ij}: it is a nonnegative precedence delay so that job j can start only d_{ij} units of time after the completion time of job i. This parameter is very general because it can represent the normal precedence constraints, by replacing $i \prec j$ with $d_{ij} = 0$, release dates, by adding a dummy 0-processing time vertex v and by setting $d_{vj} = r_j$, and also delivery times.

For the problem of minimizing the weighted sum of completion times $\sum w_j C_j$ subject to precedence delay constraints, we define an LP relaxation:

$$\min \quad \sum_{j \in N} w_j C_j$$

subject to

$$(a) \quad C_j \geq p_j \qquad \text{for all } j \in N$$

$$(b) \quad C_j \geq C_i + d_{ij} + p_j \qquad \text{for all } (i, j) \in A$$

$$(c) \quad \sum_{j \in S} p_j C_j \geq \frac{1}{2m} \left(\sum_{j \in S} p_j \right)^2 + \frac{1}{2} \sum_{j \in S} p_j^2 \qquad \text{for all } S \subseteq N.$$

Constraints (a) impose trivial lower bounds on the completion times of the jobs, constraints (b) insert the ordinary precedence constraints, while (c) were introduced in [39], for the m machines case. We show now the feasibility.

Lemma 1.23 *Given the completion times of jobs in any single machine feasible schedule we have*

$$\sum_{j \in S} p_j C_j \geq \frac{1}{2} \left(\sum_{j \in S} p_j^2 + \left(\sum_{j \in S} p_j \right)^2 \right) \qquad \textit{for all } S \subseteq N.$$

Proof We can suppose, in case by reindexing, that the jobs are ordered by completion time, so $C_1 \leq \cdots \leq C_n$. Let $S = N$, clearly we have that $C_j \geq \sum_{k \leq j} p_k$ and multiplying by p_j and summing over j it follows

$$\sum_{j=1}^{n} p_j C_j \geq \sum_{j=1}^{n} p_j \sum_{k=1}^{j} p_k = \frac{1}{2}\left(\sum_{j \in N} p_j^2 + \left(\sum_{j \in N} p_j\right)^2\right).$$

For any other set S, the jobs in S are feasibly scheduled by the schedule $\{1, \ldots, n\}$ just by ignoring the other jobs. So we may see S as the new entire set of jobs and then we apply the previous argument.

Proposition 1.10 *Given a feasible schedule on m parallel machines, the completion times satisfy constraints* (*c*).

Proof Let S_i be the set of jobs processed on machine i, for $i = 1, \ldots, m$. From Lemma 1.23 we have

$$\sum_{j \in S} p_j C_j = \sum_{i=1}^{m} \sum_{j \in S_i} p_j C_j \geq \sum_{i=1}^{m} \frac{1}{2}\left(\sum_{j \in S_i} p_j^2 + \left(\sum_{j \in S_i} p_j\right)^2\right) =$$

$$= \frac{1}{2}\sum_{j \in S} p_j^2 + \frac{1}{2}\sum_{i=1}^{m}\left(\sum_{j \in S_i} p_j\right)^2 \geq \frac{1}{2}\sum_{j \in S} p_j^2 + \frac{1}{2m}\left(\sum_{j \in S} p_j\right)^2.$$

So given any feasible solution C^{LP} of this linear program we can, applying the list scheduling algorithm, define the schedule H, with completion time C^H. We will study the relation between the completion time of jobs, so between C_j^H and C_j^{LP}. Let us define the LP midpoints as $M_j^{LP} := C_j^{LP} - \frac{p_j}{2}$ and assume they define the list L that will be used in the job-driven list scheduling.

Theorem 1.20 *Let M_j^{LP} be the midpoint of job j in the LP relaxation. Let H be the schedule obtained by running the job-driven list scheduling algorithm in order of LP midpoints, and denote by s_j^H the starting time of job j in the schedule H. Then it follows*

$$s_j^H \leq 4\, M_j^{LP} \text{ for all } j \in N. \tag{37}$$

Proof Let the jobs be indexed, in case by reindexing, in order of LP midpoint, i.e., $M_1^{LP} \leq M_2^{LP} \leq \cdots \leq M_n^{LP}$. Let j be a fixed job and suppose the algorithm runs until job j is considered. We can denote by μ the total time in $[0, s_j^H)$ where all the machines are busy. It follows, since only jobs $i = 1, \ldots, j-1$ have been processed, that $\mu \leq \sum_{i=1}^{j-1} \frac{p_i}{m}$. Then, in order to prove Eq. (37), we need to show:

$$(i)\ \frac{1}{m}\sum_{i=1}^{j-1} p_i \le 2M_j^{LP} \qquad \text{and} \qquad (ii)\ s_j^H - \mu \le 2M_j^{LP},$$

where we can denote $s_j^H - \mu =: \lambda$.

(i) : first we need to reformulate (c); for each $S \subseteq N$ this constraint is equivalent to

$$\sum_{i\in S} p_i M_i \ge \frac{1}{2m}\left(\sum_{i\in S} p_i\right)^2 .$$

By the reindexing, for each $i < j$, we have $M_i^{LP} \le M_j^{LP}$, hence

$$\left(\sum_{i=1}^{j-1} p_i\right) M_j^{LP} \ge \sum_{i=1}^{j-1} p_i M_i^{LP} \ge \frac{1}{2m}\left(\sum_{i=1}^{j-1} p_i\right)^2 .$$

So, after dividing by $\sum_{i=1}^{j-1} p_i$ it follows $\sum_{i=1}^{j-1} p_i \le 2m \cdot M_j^{LP}$.

(ii) : we want to prove $\lambda \le 2M_j^{LP}$. Let q be the number of maximal intervals in the schedule H such that at least a machine is idle in $\left[0, s_j^H\right)$. If we denote this idle intervals by $[b_h, e_h)$ for $h = 1, 2, \ldots, q$, we have $0 \le b_1 < e_1 < b_2 < \cdots < b_q < e_q \le s_j^H$. So it follows $\lambda = \sum_{h=1}^{q}(e_h - b_h)$ and all machines are busy in the complementary intervals. We know that $[j]$ is the set of jobs with LP midpoint not greater than M_j^{LP}. We consider now the directed graph G^j with vertices the jobs $[j]$ and with tight precedence constraints, so

$$G^j = \left([j], A^j\right) := \left([j], \{(k,i) \in A : k, i \in [j] \text{ and } C_i^H = C_k^H + d_{ki} + p_i\}\right).$$

If $e_q > 0$ then there is a job, namely $x(q)$, starting at time e_q ($x(q)$ can be either j). So there is a job $x(q)$ such that $s_{x(q)}^H = e_q$ and, since $x(q) \in [j]$, we have $M_{x(q)}^{LP} \le M_j^{LP}$. Let $v_1, \ldots, v_z = x(h)$ be a maximal path in A^j ending at $x(h)$, with $h = q$. We repeat this process for decreasing value of h. By the maximality of the path and by $e_q > 0$, job v_1 must have started in a busy interval $[e_g, b_{g+1})$, hence it follows $e_g < s_{v_1}^H < b_{g+1}$ for some busy interval $[e_g, b_{g+1})$ with $b_{g+1} < e_q$, otherwise job v_1 would have started before. So we have

$$e_h - b_{g+1} \le s_{v_z}^H - s_{v_1}^H = \sum_{i=1}^{z-1}\left(s_{v_{i+1}}^H - s_{v_i}^H\right) = \sum_{i=1}^{z-1}\left(p_{v_i} + d_{v_i v_{i+1}}\right). \tag{38}$$

The constraints (b) imply, for all $i = 1, 2, \ldots, z-1$, that

$$M^{LP}_{v_{i+1}} \geq M^{LP}_{v_i} + \frac{p_{v_i}}{2} + d_{v_i v_{i+1}} + \frac{p_{v_{i+1}}}{2}. \tag{39}$$

Hence, Eqs. (38) and (39) together give

$$M^{LP}_{x(h)} - M^{LP}_{v_1} \geq \frac{1}{2}\sum_{i=1}^{z-1}\left(p_{v_i} + d_{v_i v_{i+1}}\right) \geq \frac{1}{2}(e_h - b_{g+1})$$

If $e_g > 0$, then there exists a job $x(g)$ such that $s^H_{x(g)} = e_g$. By the indices of the jobs in the list we have that $s^H_{x(g)} < s^H_{v_1}$ implies $M^{LP}_{x(g)} \leq M^{LP}_{v_1}$, hence

$$M^{LP}_{x(h)} - M^{LP}_{x(g)} \geq \frac{1}{2}(e_h - b_{g+1}). \tag{40}$$

Since $s_{x(g)} = e_g$ (for the case $e_g > 0$) it follows that there is a $k \in [j]$ such that $(k, x(g)) \in A^j$ and $s^H_k < e_g = s^H_{x(g)}$. We can repeat the above procedure with $h = g$, and $x(h) = x(g)$. The whole process will terminate by the fact that at each step we have $g < h$. So we generated a decreasing sequence of indices $q = y(1) > \cdots > y(q') = 0$, in such a way that the intervals $[b_{y(i+1)+1}, e_{y(i)}]$ contain every idle interval. Then, if we add the inequalities (40) it follows:

$$\lambda \leq \sum_{i=1}^{q'-1}(e_{y(i)} - b_{y(i+1)+1}) \leq 2\,(M^{LP}_{x(y(1))} - M^{LP}_{x(y(q'))}) \leq 2\,(M^{LP}_j - 0) = 2\,M^{LP}_j. \tag{41}$$

Corollary 1.5 *Let us denote by C^{LP} the optimal solution of the LP relaxation defined by the constraints (a), (b), (c) for the scheduling problem $P|d_{ij}|\sum w_j C_j$, by C^H the solution of the job-driven list scheduling algorithm using the LP midpoints list and by C^* an optimum schedule, then*

$$\sum_{j\in N} w_j C^H_j \leq 4 \sum_{j \in N} w_j C^*_j, \tag{42}$$

and the bound is tight.

Proof The first part of the theorem, namely Eq. (42), is obvious by Theorem 1.20. For the tightness we show now an example.

Suppose we have $m \geq 2$ parallel machines and m sets of jobs J_h, for $h = 1, \ldots, m$ with $m+1$ jobs each: one medium job called $a(h)$ and m small jobs called $b_g(h)$, $g = 1, \ldots, m$. We have $p_{a(h)} = 2^{h-1-1}$, $p_{b_g(h)} = 0$, and $a(h) \prec b_g(h)$. We also have m jobs $u(i)$, for $i = 1, \ldots, m$, with $p_{u(i)} = 1$, a job j_{n-1} with processing time

$p_{n-1} = \frac{2}{m}$ and a job j_n with $p_n = 0$, such that $a(m) \prec j_{n-1} \prec j_n$. We can see that in total $N = (m+1)^2 + 1$ there are jobs.

If m is big enough the constraints $(a), (b), (c)$ define a feasible LP solution such that $C_j^{LP} = p_{a(h)} = 2^{h-1-m}$ for $j \in J_h$ and $h = 1, \ldots, m$, $C_{u(i)}^{LP} = 1 + \frac{2}{m}$, for $i = 1, \ldots, m$ and $C_{n-1}^{LP} = C_n^{LP} = M_n^{LP} = p_{a(m)} + p_{n-1} = \frac{1}{2} + \frac{2}{m}$. So the list in order of midpoints is given by $J_1, J_2, \ldots, J_m$, where each of them processes before the medium job, then job j_{n-1}, after the unit jobs $u(i)$ for $i = 1, \ldots, m$ and at the end job j_n. Then the algorithm, using a list in order of LP midpoints, works as follows:

- first it processes $J_1, J_2, \ldots J_m$, where for $k = 1, \ldots, m$:
  - a machine processes job a_k with $C_{a_k}^H = \sum_{i=1}^k p_{a(i)} = 2^{h-m} - 2^{-m}$
  - jobs $b_k(h)$, for $h = 1, \ldots, m$, are processed on the m machines with $C_{b_k(h)}^H = C_{a_k}^H$,
- it processes job j_{n-1} and jobs $u(i)$, $i = 1, \ldots, m-1$ concurrently on the m machines,
- it processes $u(m)$ on the same machine as j_{n-1}
- it processes job j_n, starting at time $s_n^H = 2 - 2^{-m}$

Let suppose now $w_n = 1$ and all the other weights are 0, then the optimum solution is $\sum w_j C_j^* = w_n(p_{a(m)} + p_{n-1} + p_n) = \frac{1}{2} + \frac{2}{m}$, while the LP midpoints list produces a solution with value $\sum w_j C_j^H = 2 - 2^{-m}$. Hence, for m sufficiently large we have $s_n^H = 2 - 2^{-m} \approx 2 + \frac{8}{m} = 4 \cdot M_n^{LP}$. So the bound is tight.

10 Conclusion

We considered some scheduling problems that arise in a wide range of applications, and we shown some possible algorithms and their performances. These algorithms used two different concepts: the LP relaxation of an integer program and the α-point scheduling, by a generic conversion algorithm.

Despite the work done on these scheduling problems, there are still some open questions or conjectures that have to be proven.

In Sect. 5, for the problem $1|r_j|\sum w_j C_j$, we presented an approximation algorithm with guarantee 1.6853, but the actual lower bound of 1.5819 is given in Sect. 5.5, so it would be interesting to close this gap.

Another interesting problem regards the problem studied in Sect. 7, namely $1|r_j, prec|\sum w_j C_j$. In a recent paper of Skutella, [41], was found a new performance guarantee of $\frac{\sqrt{e}}{\sqrt{e-1}} \approx 2.54$, that it is hardly the last word for this problem. By some nonapproximability results it seems somewhat unlikely to achieve a performance ratio strictly better than 2. Hence Skutella formulated the conjecture that, for any $\epsilon > 0$, there is a $(2+\epsilon)$-approximation algorithm for $1|r_j, prec|\sum w_j C_j$.

We also tried to use the concept of α-point for scheduling problems with parallel machines, the big problem, here, arises with the precedence constraints. In fact, while for $P|r_j|\sum w_j C_j$ the best approximation algorithm has a guarantee of 2, for the same problem with precedences, namely $P|prec, r_j|\sum w_j C_j$, the best-known performance is 4, and it is reached by an intricate algorithm which uses a fixed $\alpha = \frac{1}{2}$. We think that, with a better understanding of the α-point over parallel machines, it is possible to improve the best current bound, by modifying the α-conversion algorithm for the case with parallel machines and precedence constraints.

Our approach to Scheduling Theory was the survey Scheduling Algorithms of Karger et al. [23], where the authors, after the definition of Scheduling, wrote:

> *the practice of this field dates to the first time two humans contended for a shared resource and developed a plan to share it without bloodshed*

hence we hope that this work will head to less bloodshed.

Acknowledgements I would like to thank my supervisor, Professor Kis Tamás, for all the time that he spent with me. He was my guide in this work and I am really grateful to him that he introduce me to Scheduling Theory. But, first of all, I would like to thank my Mum and my Dad, my two brothers and all those people that I consider my family, because they made me the person who I am.

References

1. Ambühl C, Mastrolilli M (2006) Single machine precedence constrained scheduling is a vertex cover problem. In: European symposium on algorithms. Springer, Heidelberg, pp 28–39
2. Ambühl C, Mastrolilli M, Mutsanas N, Svensson O (2011) On the approximability of single-machine scheduling with precedence constraints. Math Oper Res 36(4):653–669
3. Baker KR (1974) Introduction to sequencing and scheduling. Wiley, New York
4. Balas E (1985) On the facial structure of scheduling polyhedra. In: Cottle RW (ed) Mathematical programming essays in honor of George B dantzig part I, pp. 179–218
5. Bansal N, Khot S (2009) Optimal long code test with one free bit. In: 2009 50th annual IEEE symposium on foundations of computer science, FOCS'09. IEEE, pp 453–462
6. Belouadah H, Posner ME, Potts CN (1992) Scheduling with release dates on a single machine to minimize total weighted completion time. Discrete Appl Math 36(3):213–231
7. Bruno J, Coffman EG Jr, Sethi R (1974) Scheduling independent tasks to reduce mean finishing time. Commun ACM 17(7):382–387
8. Chakrabarti S, Phillips CA, Schulz AS, Shmoys DB, Stein C, Wein J (1996) Improved scheduling algorithms for minsum criteria. In: International colloquium on automata, languages, and programming. Springer, Heidelberg, pp 646–657
9. Chekuri C, Khanna S (2004) Approximation algorithms for minimizing average weighted completion time. In: Handbook of scheduling: algorithms, models, and performance analysis. Chapman and Hall/CRC
10. Chekuri C, Motwani R, Natarajan B, Stein C (2001) Approximation techniques for average completion time scheduling. SIAM J Comput 31(1):146–166
11. Cormen TH, Leiserson CE, Rivest RL (1989) Introduction to algorithms
12. Correa JR, Schulz AS (2005) Single-machine scheduling with precedence constraints. Math Oper Res 30(4):1005–1021
13. Dyer ME, Wolsey LA (1990) Formulating the single machine sequencing problem with release dates as a mixed integer program. Discrete Appl Math 26(2–3):255–270
14. Goemans MX (1996) A supermodular relaxation for scheduling with release dates. In: International conference on integer programming and combinatorial optimization, Springer, Heidelberg, pp 288–300
15. Goemans MX (1997) Improved approximation algorithms for scheduling with release dates. Technical Report, Association for Computing Machinery, New York, NY (United States)
16. Goemans MX, Queyranne M, Schulz AS, Skutella M, Wang Y (2002) Single machine scheduling with release dates. SIAM J Discrete Math 15(2):165–192

N. Gusmeroli, *Machine Scheduling to Minimize Weighted Completion Times*, SpringerBriefs in Mathematics, https://doi.org/10.1007/978-3-319-77528-9

17. Graham RL, Lawler EL, Lenstra JK, Kan AR (1979) Optimization and approximation in deterministic sequencing and scheduling: a survey. Ann Discrete Math 5:287–326
18. Hall LA (1996) Approximation algorithms for scheduling. In: Approximation algorithms for NP-hard problems. PWS Publishing Co., pp 1–45
19. Hall LA, Schulz AS, Shmoys DB, Wein J (1997) Scheduling to minimize average completion time: off-line and on-line approximation algorithms. Math Oper Res 22(3):513–544
20. Hazewinkel M (2013) Encyclopaedia of mathematics: volume 6: subject index author index. Springer Science & Business Media
21. Hochbaum DS, Shmoys DB (1988) A polynomial approximation scheme for scheduling on uniform processors: Using the dual approximation approach. SIAM J Comput 17(3):539–551
22. Horn W (1974) Some simple scheduling algorithms. Naval Res Logistics (NRL) 21(1):177–185
23. Karger D, Stein C, Wein J (2010) Scheduling algorithms. In: Algorithms and theory of computation handbook. Chapman & Hall/CRC, p 20
24. Labetoulle J, Lawler EL, Lenstra JK, Kan AR (1982) Preemptive scheduling of uniform machines subject to release dates. Stichting Mathematisch Centrum Mathematische Besliskunde (BW 166/82):1–20
25. Lawler EL (1978) Sequencing jobs to minimize total weighted completion time subject to precedence constraints. Ann Discrete Math 2:75–90
26. Lawler EL, Lenstra JK, Kan AHR, Shmoys DB (1993) Sequencing and scheduling: algorithms and complexity. Handbooks Oper Res Manag Sci 4:445–522
27. Lenstra JK, Rinnooy Kan A (1978) Complexity of scheduling under precedence constraints. Oper Res 26(1):22–35
28. Lenstra JK, Kan AR, Brucker P (1977) Complexity of machine scheduling problems. Ann Discrete Math 1:343–362
29. Motwani R, Raghavan P (2010) Randomized algorithms. Chapman & Hall/CRC
30. Phillips C, Stein C, Wein J (1995) Scheduling jobs that arrive over time. In: Workshop on Algorithms and Data Structures. Springer, Heidelberg, pp 86–97
31. Posner ME (1986) A sequencing problem with release dates and clustered jobs. Manag Sci 32(6):731–738
32. Queyranne M (1987) Polyhedral approaches to scheduling problems. In: Seminar presented at Rutcor, Rutgers University
33. Queyranne M (1993) Structure of a simple scheduling polyhedron. Math Program 58(1):263–285
34. Queyranne M, Schulz AS (1995) Scheduling unit jobs with compatible release dates on parallel machines with nonstationary speeds. In: International Conference on Integer Programming and Combinatorial Optimization. Springer, pp 307–320
35. Queyranne M, Schulz AS (2006) Approximation bounds for a general class of precedence constrained parallel machine scheduling problems. SIAM J Comput 35(5):1241–1253
36. Schulz A, Skutella M (1997) Random-based scheduling new approximations and LP lower bounds. In: Randomization and approximation techniques in computer science, pp 119–133
37. Schulz AS (1996) Scheduling to minimize total weighted completion time: performance guarantees of LP-based heuristics and lower bounds. In: International conference on integer programming and combinatorial optimization. Springer, pp 301–315
38. Schulz AS, Skutella M (2002) The power of α-points in preemptive single machine scheduling. J Scheduling 5(2):121–133
39. Schulz AS et al. (1996) Polytopes and scheduling. PhD thesis, Technische Universität Berlin
40. Skutella M (1998) Approximation and randomization in scheduling. Cuvillier
41. Skutella M (2016) A 2.542-approximation for precedence constrained single machine scheduling with release dates and total weighted completion time objective. Oper Res Lett 44(5):676–679
42. Smith WE (1956) Various optimizers for single-stage production. Naval Res Logistics Q 3(1–2):59–66
43. Stein C, Wein J (1997) On the existence of schedules that are near-optimal for both makespan and total weighted completion time. Oper Res Lett 21(3):115–122

44. Torng E, Uthaisombut P (1999) Lower bounds for srpt-subsequence algorithms for nonpreemptive scheduling. In: Symposium on discrete algorithms: proceedings of the tenth annual ACM-SIAM symposium on discrete algorithms, vol 17, pp 973–974
45. Uma R, Wein J (1998) On the relationship between combinatorial and LP-based approaches to NP-hard scheduling problems. In: International conference on integer programming and combinatorial optimization. Springer, pp 394–408
46. Wolsey L (1985) Mixed integer programming formulations for production planning and scheduling problems. In: Invited talk at the 12th international symposium on mathematical programming. Massachusetts Institute of Technology, Cambridge, MA

CPSIA information can be obtained
at www.ICGtesting.com
Printed in the USA
BVHW01s0237160518
516401BV00003B/47/P

9 783319 775272